AF413407

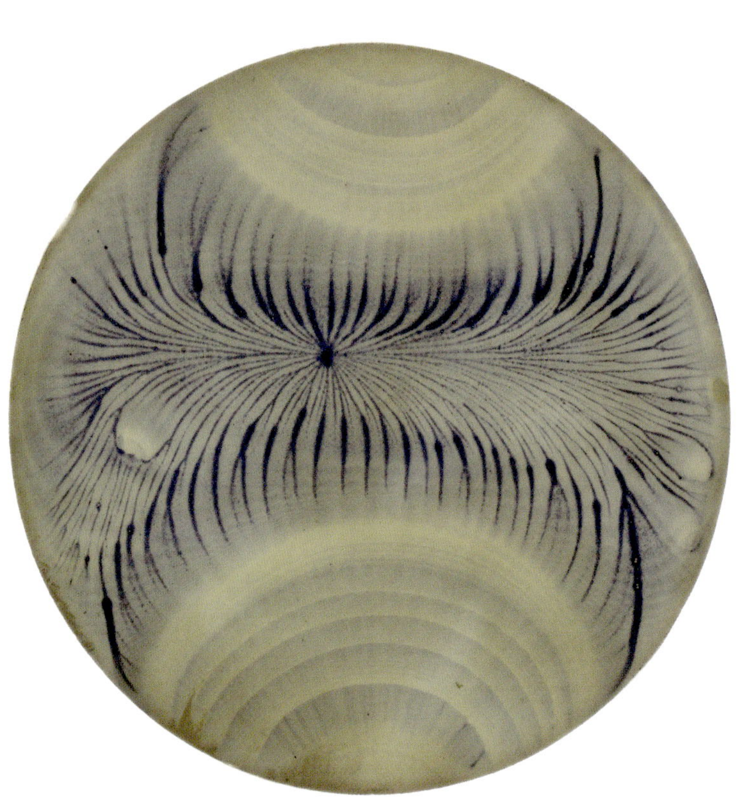

MARGARET WATTS HUGHES

SOUND MAY BE SEEN

THE FURTHER READING LIBRARY

CHRISTINE BURGIN BOOKS

SERIES EDITORS: CHRISTINE BURGIN AND ANDREW LAMPERT

Published by Christine Burgin Books
Ghent, New York
© 2025 Christine Burgin and Andrew Lampert
"Visible Sound" © 2025 Rob Mullender-Ross

Front cover and pages 2, 14, 40–41, 45–75, 84, and 86–87: Louis Porter;
page 9: courtesy Cyfarthfa Castle Museum and Art Gallery; pages 13 and
21–31: Rob Mullender-Ross; page 19 and back cover: Daniel R. Wilson

ISBN: 978-0-9976456-6-8

Published in 2025

www.furtherreadinglibrary.com

Special thanks to Michelle Lewis and to the Cyfarthfa Castle Museum and
Art Gallery, Wales, UK

Designed and typeset in Caslon by Laura Lindgren
Image preparation by Jason Burch
Front cover lettering by Craig Pettman
Printed on 115 gsm Kasadaka White
Printed in China

FRONTISPIECE: *Voice Figure, ca. 1885–1904, paint on glass, 5 × 5 in.
(12.7 × 12.7 cm)*

CONTENTS

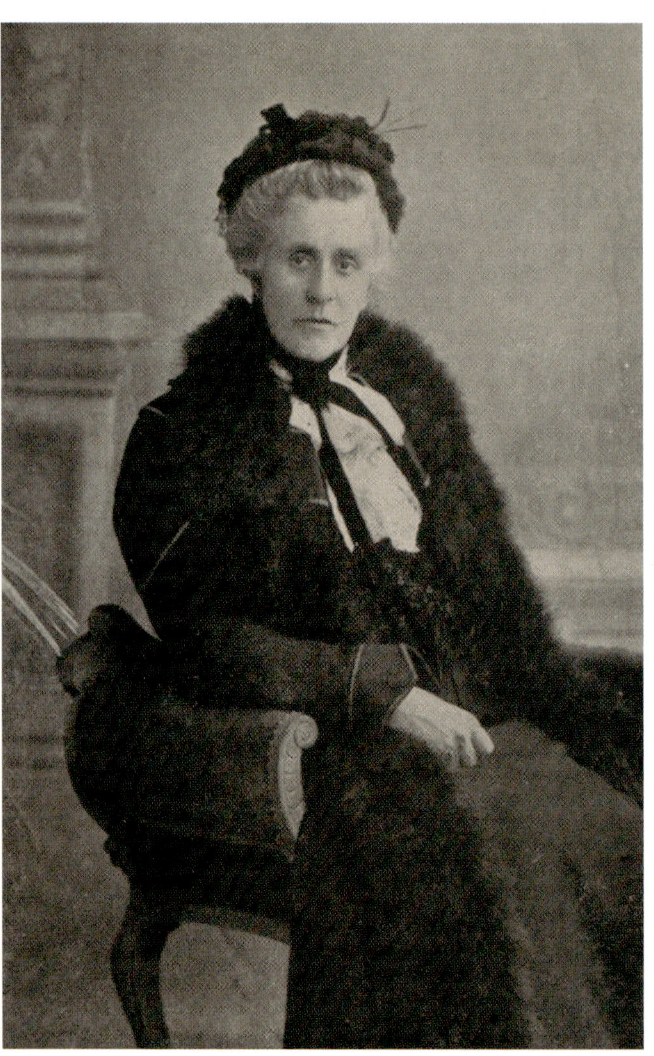

VISIBLE SOUND

Rob Mullender-Ross

Is the violet the form of a sound,
the sunflower a materialized musical note?
—"Visible Sound," *The Lancaster Daily*
Intelligencer, Pennsylvania, May 4, 1891

In the late nineteenth century, the acclaimed Welsh soprano and philanthropist Margaret Watts Hughes invented an apparatus that made it possible for sound to be seen. She named this device the eidophone, and with it she created a stunning array of visual transcriptions called "Voice Figures." Scientists, theosophists, and the public at large were fascinated by her experiments with acoustics and her remarkable ability to sing into existence flowers, ferns, landscapes, and colorful abstractions. Black-and-white reproductions of her works were internationally published and discussed; however, it was not until a collection of "Voice Figure" glass slides came to light at Cyfarthfa Castle Museum, Merthyr Tydfil, in Wales in 2016 that their rich colors could at last be seen and appreciated. Watts Hughes's alluring images seem to have as much in

Margaret Watts Hughes as she appears in Eminent Welshmen: A Short Biographical Dictionary of Welshmen Who Have Attained Distinction from the Earliest Times to the Present *by T. R. Roberts (London: Cardiff & Merthyr Tydfil, 1908)*

common with the natural sciences as they do with art, music, and the metaphysical potential of making the invisible visible. A century and a half later the wonderment and vibrations produced by these extraordinary works continue to resonate.

Margaret (Megan) Watts was born in 1842 in Dowlais, South Wales, a village dominated by the Dowlais Ironworks, one of the UK's largest industries at the time. Watts was raised in the Congregationalist Christian tradition, of which South Wales was a stronghold in the mid-nineteenth century, and faith proved to be a driving force throughout her life. As a child she was highly attuned to sound, especially the constant and cacophonous noises emanating from the Ironworks. An exceptionally gifted singer who began performing as a soloist with her choir at the age of ten, Watts was held in high esteem within Welsh popular and religious music circles. She moved to London in 1869 to pursue studies at the Royal Academy of Music and became a regular fixture on the concert circuit. In 1872 Watts married Hugh Lloyd Hughes and became known as Margaret Watts Hughes. Her musical career continued through the early 1880s, around which time she largely stepped back from solo performing to engage more fully in charitable endeavors.

Concerned about the state of the Welsh poor in London, Watts Hughes spent time making house visits to destitute families and became directly involved with missionary work. She taught a Bible class for the girls in the choir at her church that evolved into a singing class for more than one hundred disadvantaged and unhoused children. Enthused by her religious

Children's Voices: A Collection of New Tunes, Duets, Rounds, and Anthems, for the Home and the School, Composed by Mrs. Watts-Hughes. *London: Morgan and Scott, 1881*

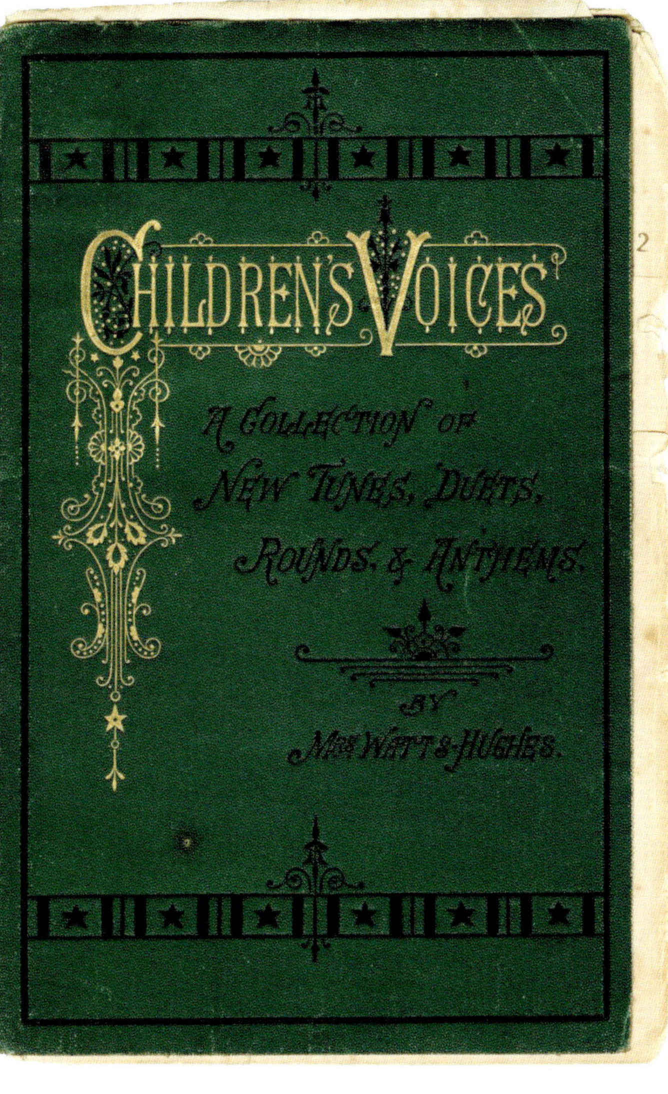

CHILDREN'S VOICES
A COLLECTION OF
NEW TUNES, DUETS,
ROUNDS, & ANTHEMS.
BY
MRS. WATTS-HUGHES.

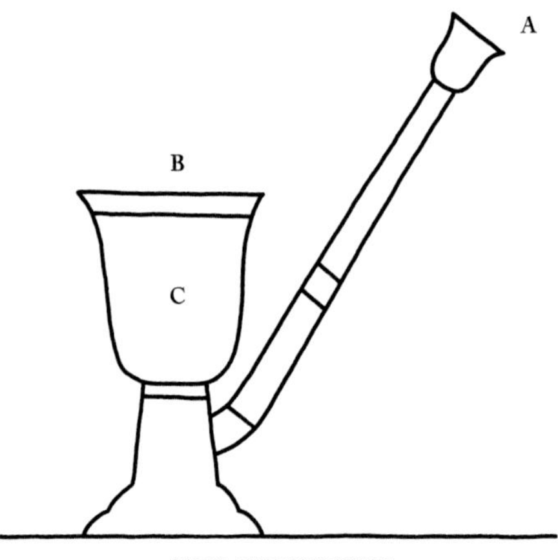

THE EIDOPHONE

"This apparatus, simple though it may be, may be said to have opened not only a new department in the domain of science, but also a new art, demanding the highest skill of the vocalist to interpret, and which appeals to human beings not by the usual way, orally, but by a new way—the visual sense.

"The illustration herewith shows one of the most convenient forms. A sustained vocal note is directed into the tube A; its vibrations strike the elastic indiarubber disc B, stretched across the mouth of the receiver C.

"If the disc be thoroughly flexible, evenly stretched, and perfectly level, the sounds waves will cause it to vibrate regularly.

"Its vibrations will vary in rapidity with the pitch of the note producing them; and they will divide the disc and subdivide the disc, creating centers of motion, and boundaries of rest, which in their turn govern the distribution and arrangement of the substance placed upon the surface of the disc, and thus form the Voice Figures."

Margaret Watts Hughes,
The Eidophone: Voice Figures *(1904)*

convictions, and supported by her wealthy husband, Watts Hughes established Mountford House, a home for destitute boys in Islington, North London. A February 2, 1889, *Daily News* article notes that "The boys whom Mrs. Hughes accommodates would have no home but for her. She wisely sends them to the Board school for their regular education, supplies the lack of a mother to them as well as a wise and large-hearted woman can, and equips them for the battle of life by instructing them in practical ways. The institution is not a large one, for Mrs. Hughes does everything by personal influence, and no substitute could supply her place."

It was also during this period that Watts Hughes's passion for singing took a decidedly scientific turn. Her extensive musical interests extended to reading about discoveries that scientists were making in the field of sound and acoustics. She was aware of the German physicist and musician Ernst Chladni's milestone experiments involving powders forming patterns on vibrating metal plates. Her brother, John Watts, wrote that she persuaded the physicist John Tyndall, the author of a popular and seminal book about sound, to demonstrate a Chladni plate for her in person. Seeing this sudden production of form from tone was revelatory for Watts Hughes, whose focus shifted from measurement to interpretation and creation.

The eidophone was invented by Watts Hughes as a tool to study the mechanics of her own voice, including volume, timbre, and pitch. She set her vocal talents to the task of defining and further refining a taxonomy of visual forms. Her initial experiments with powders used plaster of paris and lycopodium spores, then commonplace in physics demonstrations. In time, her ability to compose floral patterns and other natural shapes with sound sharpened, and she eventually progressed to using pigmented "semi-liquids" that were easier to manipulate. The eidophone would be set upon a flat surface with the vibrating diaphragm facing upward. This allowed her to "fix" these forms (or record them, if one thinks of the shapes as a form of visual music) by placing a glass plate on top of the diaphragm to take a direct "print" of the completed pattern.

Watts Hughes publicly presented her Voice Figure plates on at least a few occasions, and in 1887 was one of the first women to display her scientific work as part of the Soirées hosted by the prestigious Royal Society in London. In 1889 the novelist Emilie Barrington wrote a passionate letter to the editor of *The Spectator* after visiting Watts Hughes in Islington. She provides a fascinating description of how these works were displayed.

> Instead of blinds or curtains drawn across the lower panes of the windows, there are wonderful designs in colour; strange, beautiful things—suggesting objects in Nature, but which are certainly neither exact repetitions nor imitations of anything in Nature . . . [they] are more like, perhaps, what a dream might make out of the impressions left by Nature, perfectly drawn designs of shell-like forms, photographically precise renderings of shapes of which the exact originals were never seen by human eye on sea or land; such things as "Alice in Wonderland" might have come upon, had she tumbled down to the bottom of the sea.

A series of figures obtained using an eidophone with a disc measuring 4⅞ in. (12.4 cm) in diameter. They are numbered in the succession in which they appeared with each rise in pitch.

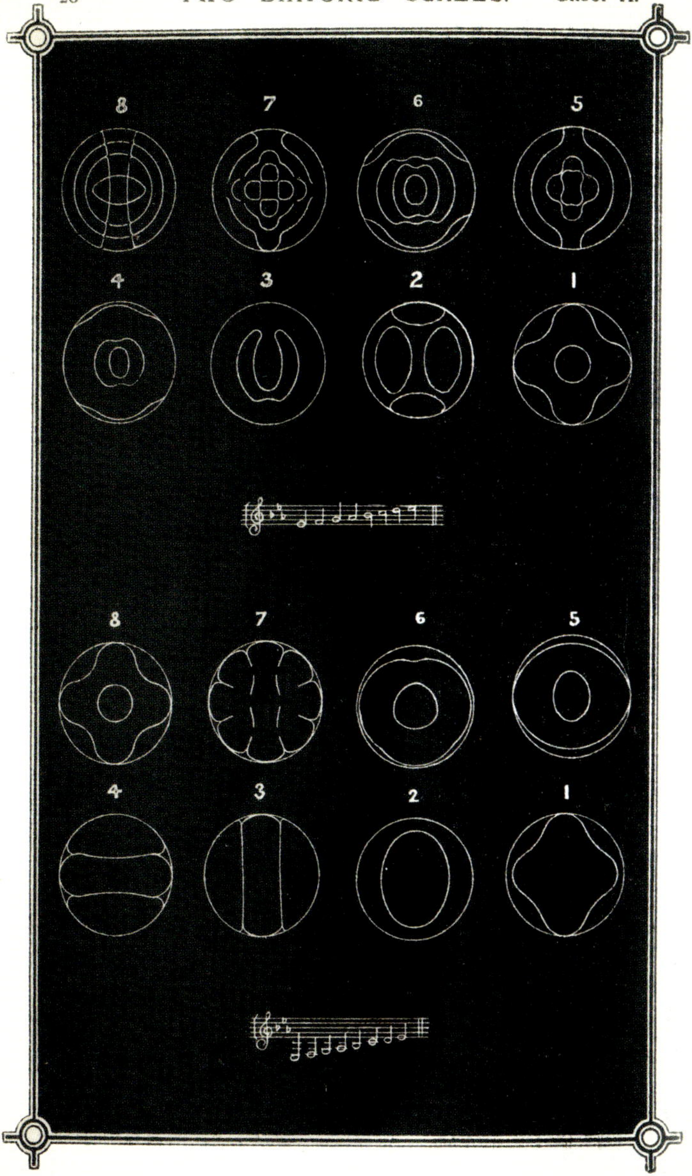

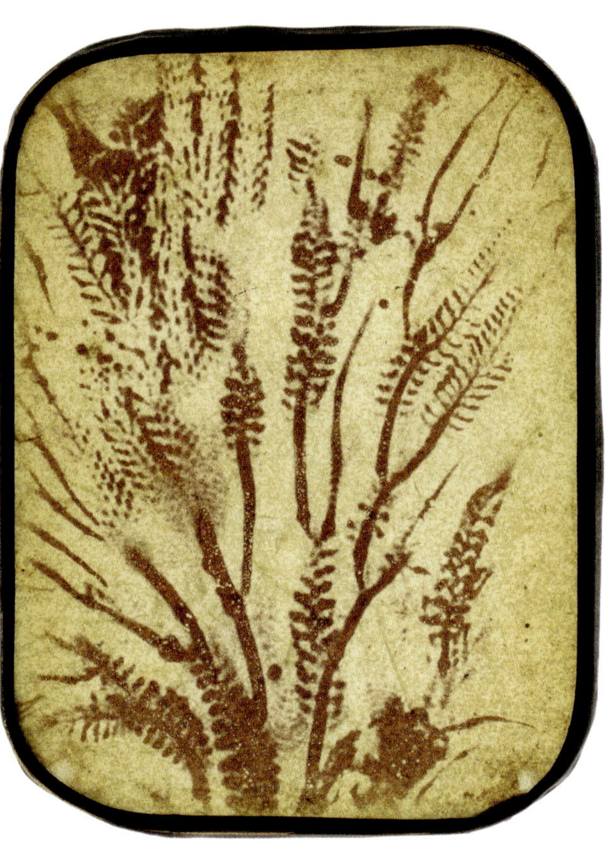

Watts Hughes announced her work to the public in 1891 in a lengthy article for the popular *Century Magazine* and in the first of what would be two editions of her book *Voice Figures*. Here, she organized her findings into a hierarchical structure moving from the "primitive" geometric to more advanced "floral" and "fern" patterns. In one revealing passage, she recounts the experience of finding "just the right kind" of sound to produce what she needs.

At first, when directing the voice against the semi-liquid mass upon the centre of the disc, there is a feeling as if some impassable barrier were encountered, and that it would be as easy to move a mountain with a push of the hand as to set that coloured heap moving by the action of a note. The next sustention seems only to confirm the first impression; but after several attempts one comes to feel that it could be done if only the right kind of sound could be employed. Persevering, the seemingly ponderous, inert mass is at last disturbed, and shows some susceptibility of control. Still continuing, it now begins to move, and ere comes under complete control, expanding in petals after every repeat crescendo. When the mass moves thus easily, the sensation of the singer is completely changed. The feeling is now as if all at once the air in the tube, in the receiver, upon the disc, and all around, were acting in concert for the singer's purpose, and had taken possession of every corner of space.

Voice Figure, ca. 1885–1904, paint on glass, 5⁹⁄₁₆ × 4¼ in. (14.1 × 10.8 cm)

The most elaborate of the Voice Figures in the Cyfarthfa Castle Museum archive (see pages 46–67) were produced using a model that was called the Hand Eidophone. She sang into the device while moving it over the surface of a glass plate coated in wet pigment; these are where one can most clearly see sound as having been imprinted. In the blue figures (usually annotated with musical pitches), the vibrating membrane leaves striations from its trailing edge as it glides across the plate, all seemingly achieved in a single, fluid movement of hand and voice. In these large plates we can see the full extent of Watts Hughes's ambitions for these methods: the singing of complex botanical formations into being. Close examination of these plates strongly resembling ferns or arborescent structures reveals a huge variety of patterns, striations, and reticulations, a true chorus of voices.

While extensive and enthusiastic press coverage in the UK and US largely focused on her scientific achievements, Watts Hughes also captured the attention of the growing Theosophical movement. Henry Steel Olcott's article in the September 1890 edition of *The Theosophist* (see pages 35–44) succinctly places the Voice Figures within the operational model of mediumistic art when declaring that these images are "drawn out of the world of invisible forms by the vibrations of Mrs Watts Hughes' glorious voice." Fifteen years later Annie Besant and C. W. Leadbeater acknowledged the work of Watts Hughes in their book *Thought Forms*, one of the most well known of all Theosophical publications.

Theosophists may have viewed the Voice Figures as manifestations from the nonphysical realm, but it is important not to lose sight of the fact that Watts Hughes was a deeply devout Christian. For her, the Voice Figures articulated the natural world in order to deepen one's appreciation of God's creation.

While acknowledging that she might be unaware of the "psychical process going on within her brain," Olcott described Watts Hughes's images as

> proof that we are living in a time of, what might be called, akasic *osmosis*—the infiltration of the astral into the physical plane of consciousness. The partition between the two worlds is proving itself as porous and penetrable as the membrane between two fluids of different densities in the familiar chemical experiment. The deeper potentialities of thought, life-force, sound, chromatics, heat and electricity, are being successfully revealed.

What scientific or spiritual proof did Watts Hughes herself find revealed in the Voice Figures? In the May 1891 *Century Magazine* article that introduced her discovery to the public, she concluded with this thought:

> . . . my experiments have been made as a vocalist, using my own voice as the instrument of investigation, and . . . the hope has come to me that these humble experiments may afford some suggestions in regard to nature's production of her own beautiful forms, and may thereby aid, in some slight degree, the revelation of yet another link in the great chain of the organized universe that, we are told in Holy Writ, took *its* shape at the voice of God.

Rob Mullender-Ross is an artist, researcher, and educator, whose multidisciplinary practice explores the incarnations and interactions of sound across sculpture, performance, installation, moving image, and text.

In this work no attempt is made to give an explanation of the phenomena referred to in the description of the various forms.

This New Science, which is given to us by one note of the human voice, aided by simple methods of operation, yields such varied and complex results that appear not only marvelous but almost miraculous. It is a science which, for explanation of its phenomena, points to the musical world as well as to the scientific world; to the vocal artist, as also to the searcher after scientific truth, and to the interpreters of those mysterious laws which underlie the visible forces of Creation.

From Margaret Watts Hughes's preface to the second edition of *Voice Figures* (*The Eidophone: Voice Figures*. London: Christian Herald Company, 1904)

VOICE FIGURES,

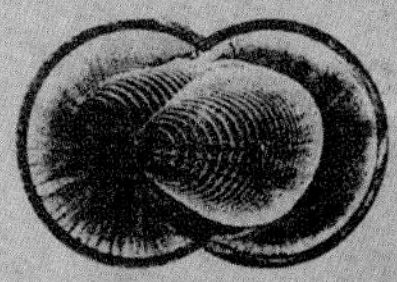

BY

Mrs. WATTS HUGHES

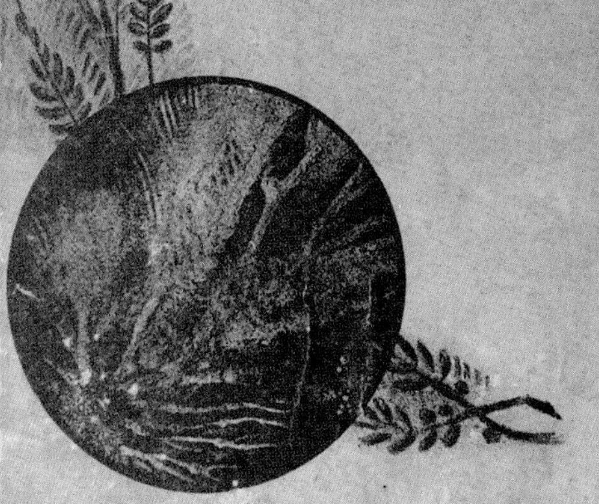

LONDON.
HAZELL, WATSON & VINEY, Ld
1. Creed Lane. Ludgate Hill. E.C.

Margaret Watts Hughes published two books about her invention of the eidophone and its applications. The first, in 1891, includes a preface by Walter Besant, who, having attended "one of Mrs. Hughes's illustrations of the connection between sound and form" the previous year, is pleased to announce that "here has been opened a door to discoveries the most interesting." Walter Besant was the brother of Annie Besant, co-author with C. W. Leadbeater of the influential Theosophical book Thought-Forms *(1905). Whereas Watts Hughes gives sound a visual form with the use of her eidophone,* Thought-Forms *describes a process in which the shapes and colors of thoughts and emotions are visualized psychically. Although their methods differ, Besant and Leadbeater acknowledged the work of Watts Hughes in their text.*

The cover on page 19 and the images that follow on pages 21–31 are from the 1891 edition of Voice Figures *by Margaret Watts Hughes, published by Hazell, Watson, and Viney Ltd., London. The cover of the 1904 edition appears on the back cover of this book.*

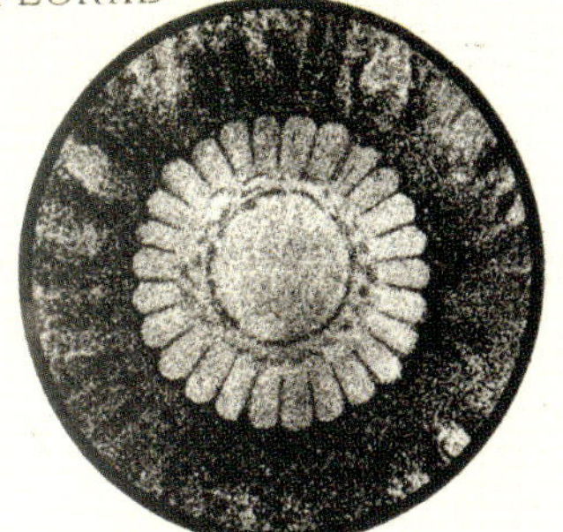

(DAISY.)

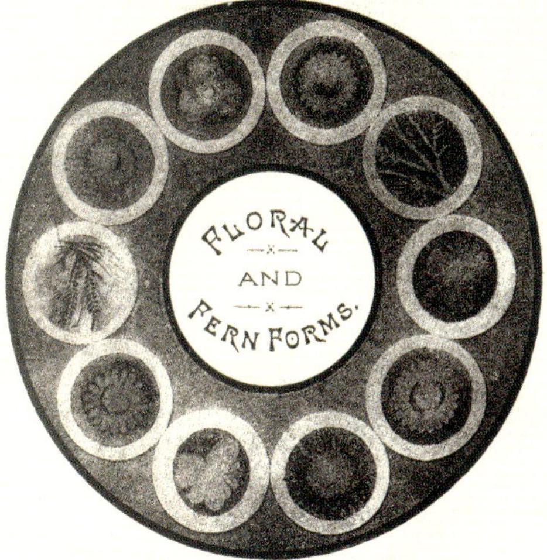

(DAISY.)

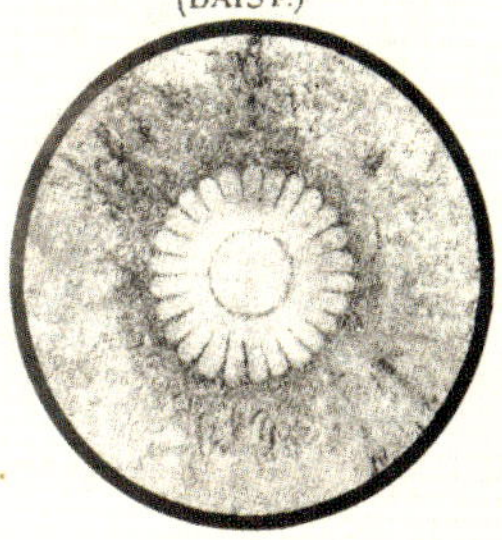

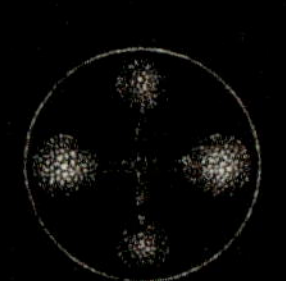 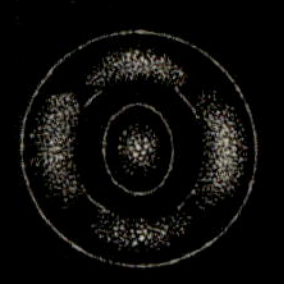 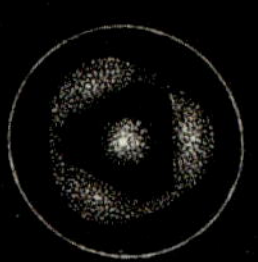 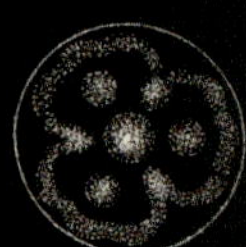

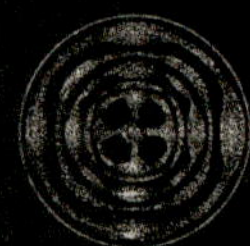 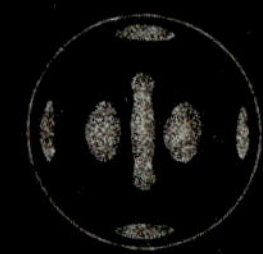 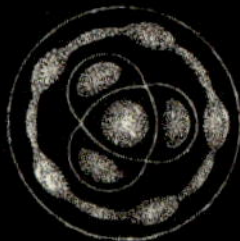 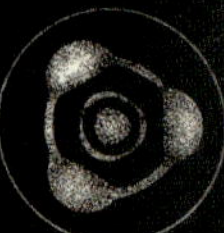

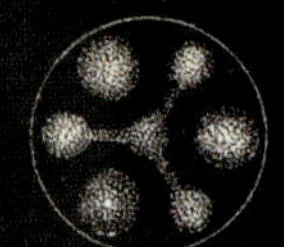 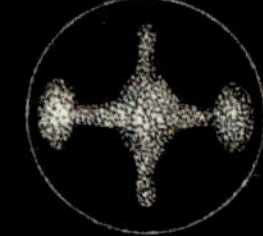 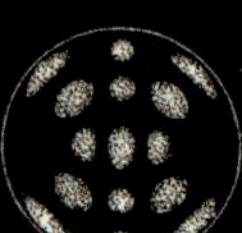 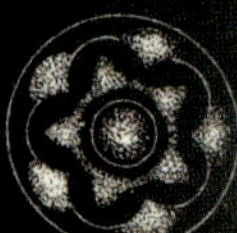

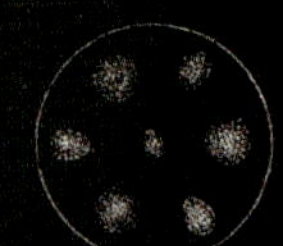 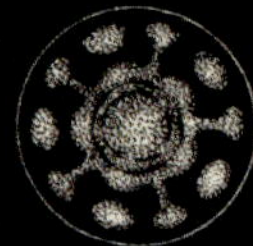 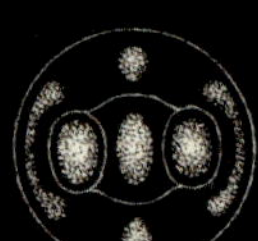 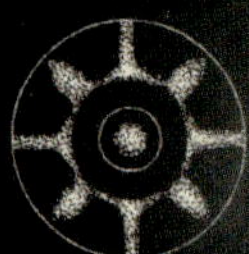

STAGE I.   PRIMITIVE FORMS.

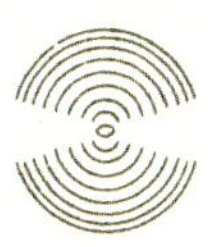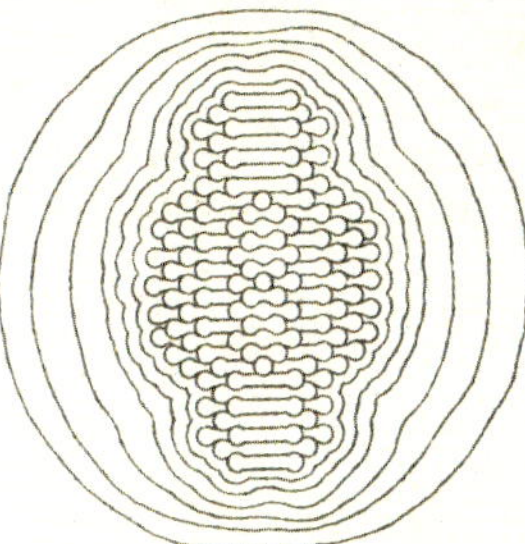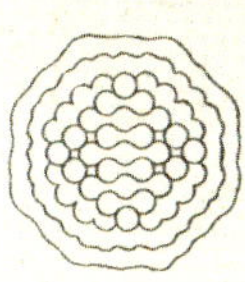

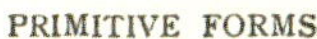

STAGE II.

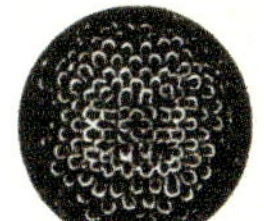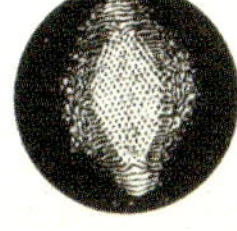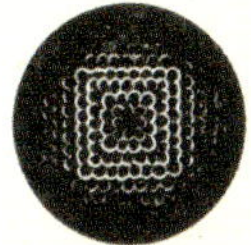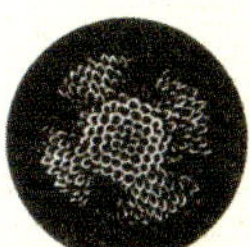

STAGE III.

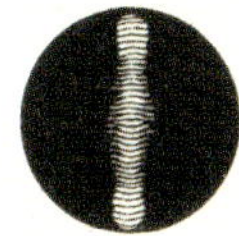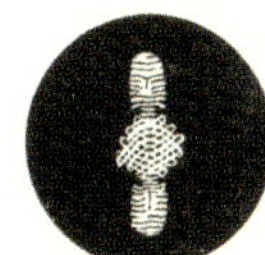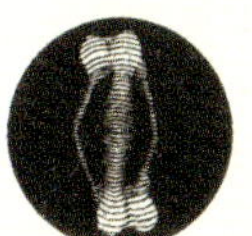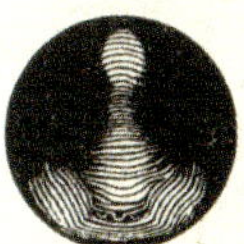

STAGE IV.

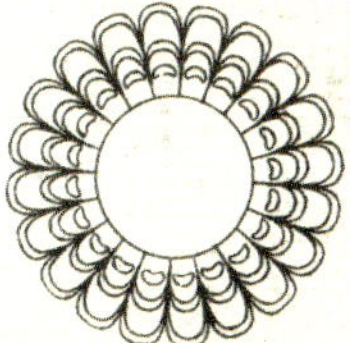

FIGURES IN PLASTER OF PARIS.

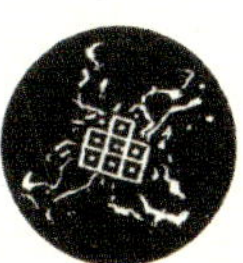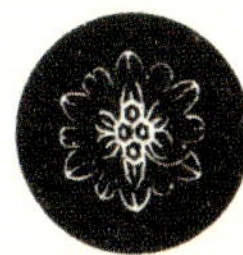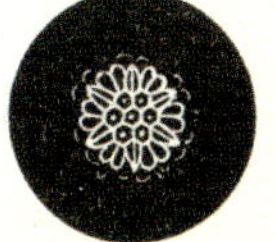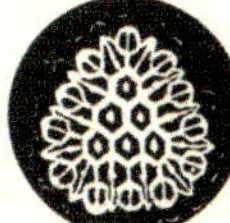

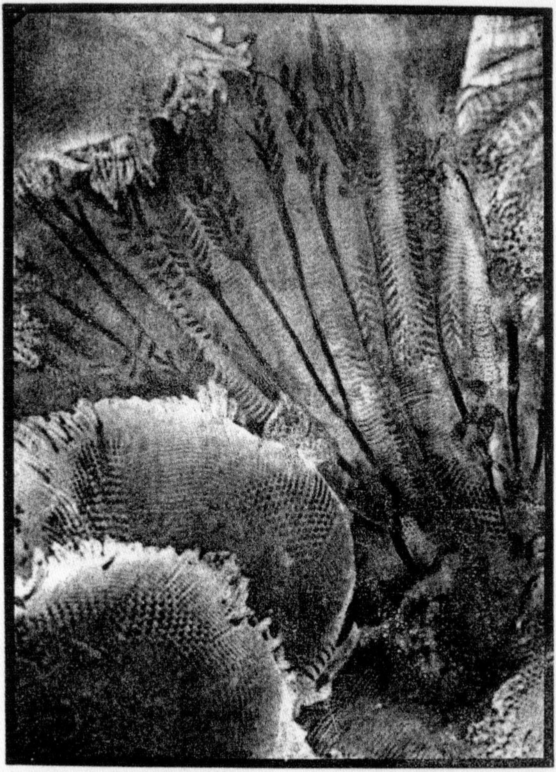

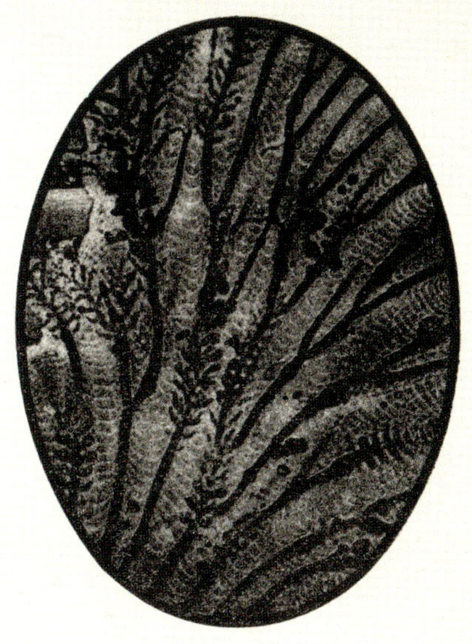

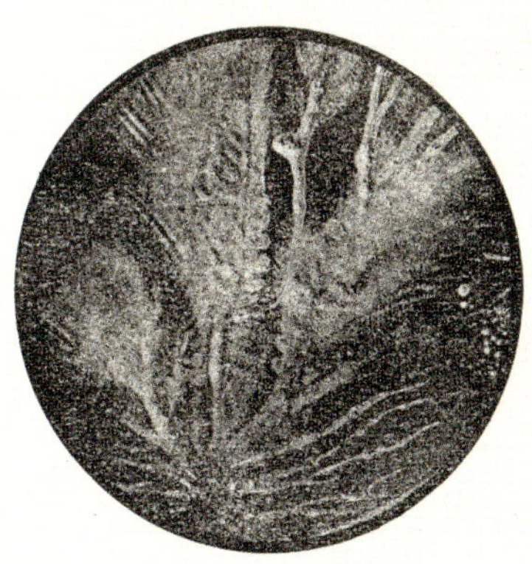

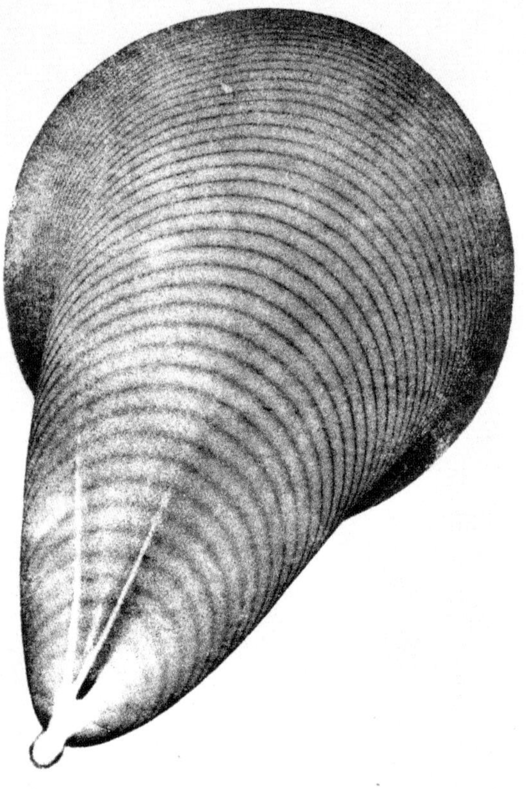

This represents the note   the figure on Plate 39 being its octave.

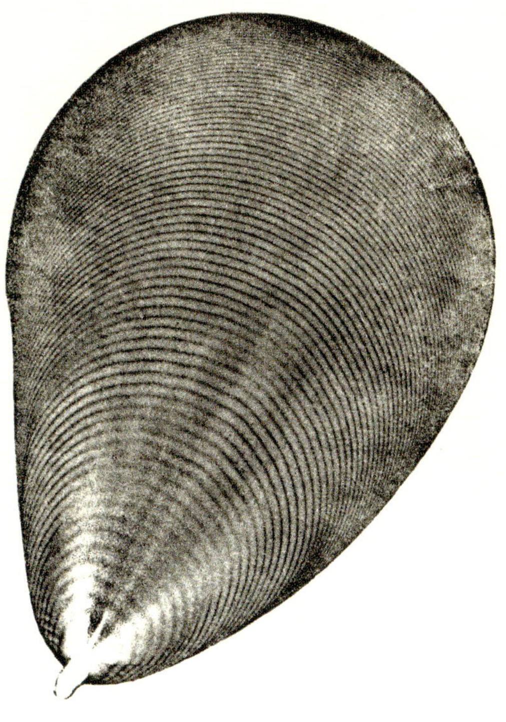

This represents the upper octave of figure 38, viz.: 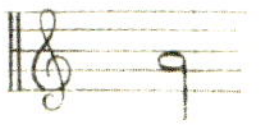

(LYCOPODIUM.)

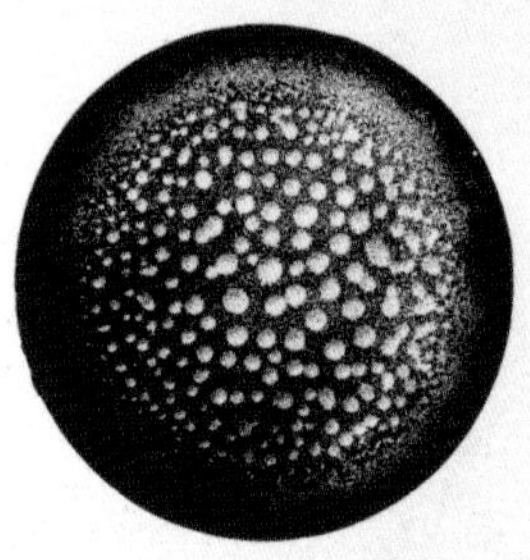

CENTRE OF MOTION.   GROUP D.

(MOIST COLOUR.)

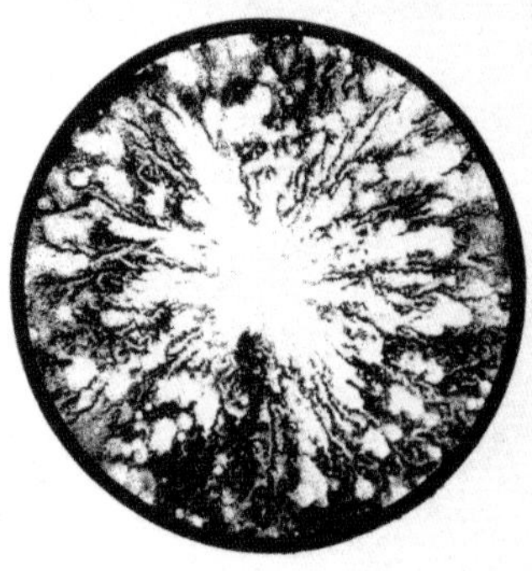

(LYCOPODIUM.)

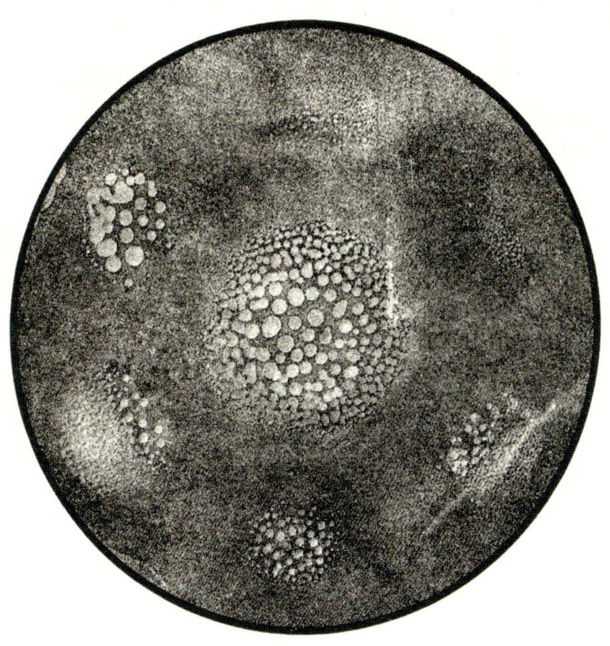

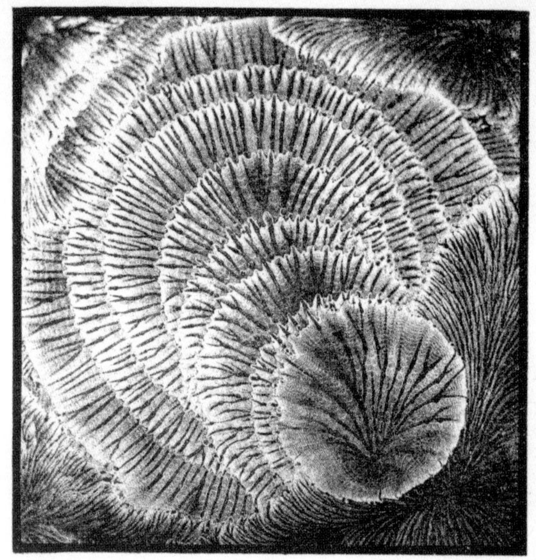

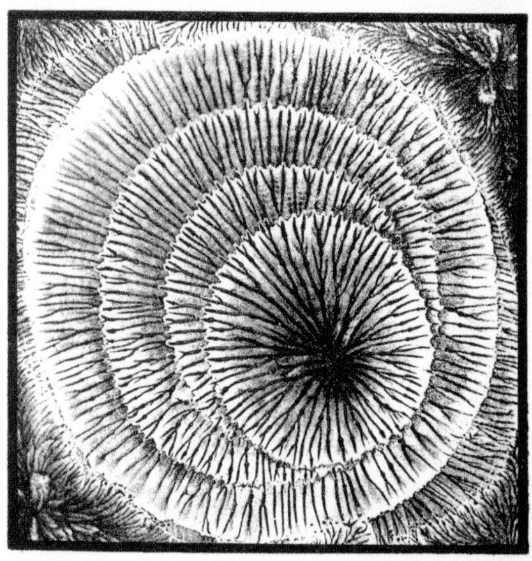

Groups X, EX, and F. IMPRESSION FIGURES & LINEAR CURVES.

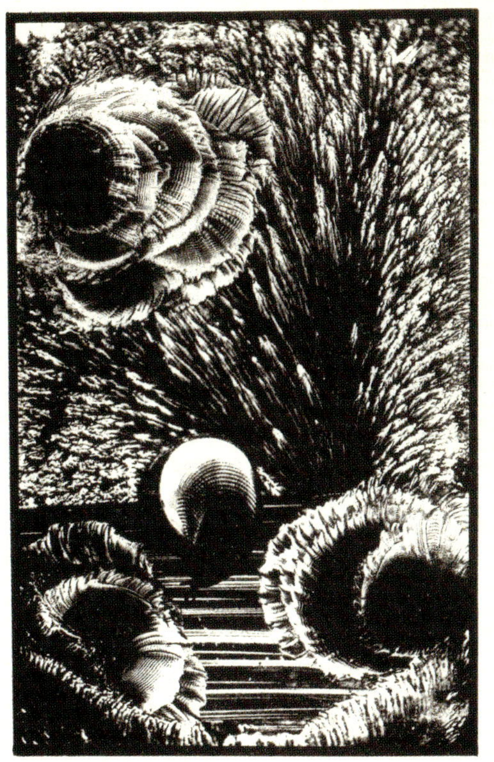

VOICE FIGURES, AND WHAT THEY ARE.

AN INTERVIEW WITH MRS. WATTS HUGHES.

THERE is quite a flutter of excitement in the world of art and of music over a wonderful discovery which confutes the proverbial saying that woman never yet made any important scientific discovery, and which at the same time seems to justify to some extent, and after twenty-five centuries, the theory of the philosopher of the golden thigh that a whole universe of harmony is around us. But Pythagoras with his mystical theories, his tapping about in the dark for a fuller explanation of the dim music which was for ever in his ear, of the seven planets forming the golden chords to the heavenly heptachord, lived only to be ridiculed by the sages of succeeding ages. The centuries swept on, and no one with a voice to speak of them to the world heard the sounds. Then, a century ago, Chladni, the natural philosopher, arose in Germany, and the world listened for a time in mute wonder and astonishment to his explanations of the laws of sound, and scientists became dimly aware of the possible uses to which sound might be put. Chladni's outline figures in sand, and his experiments on the effect of vibration on metallic and glass plates were a step further into the unknown world. Now for more than half a century Chladni's voice has been hushed, and the "philosophy of sound" seemed all but forgotten, when what science has been unable to unravel has, almost by chance, been discovered and perfected by a woman and an artist. This woman is a Welsh lady, Mrs. Watts Hughes, who gave a representative the following outline of her discoveries:—

"Are they really what you call them—voice figures; that is to say, are these marvellous pictures produced solely by the human voice and without any manual work?" I asked, after a careful survey of the figures.—"I shall be quite correct in saying," replied Mrs. Watts Hughes, "that none of them can be produced without the voice, but trees and ferns are complicated by cohesion, requiring some slight manipulation.

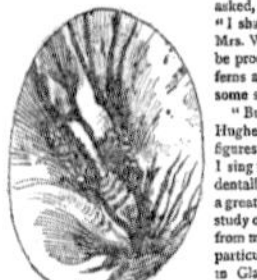

PALM.

"But tell me how you produce them, Mrs. Hughes, and how you first discovered the figures?"—"I will show you presently how I sing them, and found them out almost accidentally. They were the immediate result of a great deal of previous thought given to the study of vocal sounds. I should tell you that from my earliest childhood I have always been particularly sensitive to sound. At Dowlais, in Glamorganshire, where I lived when a child, I had the roar of the great iron-works day and night in my ears, and I heard the music in the deep thunder of the steam hammers and engines. It was perhaps this which made me interested in, and sensitive to, every kind of sound, and this interest was intensified when, later on, my musical education was begun. And the more and the longer I thought of it, the more I became convinced that the power of the human voice has never yet been realized.

"But to return to the voice figures. I had long been trying to test, as it were, the strength of individual notes of my voice by various means, and I made a great many studies of vocal sounds. While trying to discover some means by which to register visibly the vibrations of the voice, and for testing its quality and tone, I saw one day with intense surprise that the grains of sand with which I experimented formed themselves into a geometrical figure not unlike those which Chladni discovered. In fact the figures which I then produced were Chladni's figures discovered over again. I continued my investigations, and slowly and gradually discovered that by singing certain notes into the eidophone over the

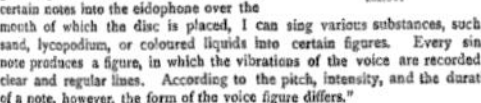

DAISY.

mouth of which the disc is placed, I can sing various substances, such as sand, lycopodium, or coloured liquids into certain figures. Every single note produces a figure, in which the vibrations of the voice are recorded by clear and regular lines. According to the pitch, intensity, and the duration of a note, however, the form of the voice figure differs."

"Then, is a certain figure always produced by singing a corresponding note?"—"The daisy and pansy and all the geometrical figures can be produced by any note of the scale; but there are many other points to be considered in the mode of production, which it would take too long to explain. Now I will show you how I work." Mrs. Hughes sat down in front of the eidophone, on which a small quantity of fine powder had been scattered. A deep, full note was sung into the tube, and immediately a miniature storm raged on the disc. Tiny clouds of dust arose, rolling and whirling about as when a hurricane sweeps over a dusty highroad. Slowly the chaos was reduced to order, and when the last vibration ceased, an accurate, clear geometrical figure lay before us, formed of the yellow powder on the dark disc.

["Now I'll change it into another figure," said Mrs. Hughes, and once again the storm began as a rich, melodious note went up through the tube. The forms produced in sand or powder are, of course, unstable, and change or vanish when the instrument is moved, but in moist colour they can be perpetuated, and the experiments with paste are, therefore, infinitely more interesting. Thus, for instance, when a daisy is to be created, the substance placed on the disc creeps together in the centre of the membrane at the command of the first note, which is obeyed as unhesitatingly as the bugle horn in the soldiers' camp. Another note follows of a different calibre, and out from the centre all round shoot small petals shaped exactly like those of the daisy. But perhaps the note has not been quite so full as it should have been, and the delicate petals are not quite as symmetrical as they ought to be. Then comes one of the most marvellous parts of the process, for once more the voice commands, and immediately they all rush back and amalgamate once again with the centre, to reappear at the next note, smooth and shapely. The process of singing waves in and waves out goes on for some time, till the last wave brings out the perfect daisy form. The size of the figures is determined by the pitch of the note and the quantity of substance to be set in motion; and as with the daisy and the pansy, so it is with all the other figures, some of which are marked with the most perfectly shaded fluted lines, representing each vibration of the voice. "To me the whole matter becomes more wonderful the longer I study it," said Mrs. Hughes, looking with thoughtful eyes on the strange beautiful forms. "There lies a whole hidden world behind these forms, which the future may perhaps reveal."

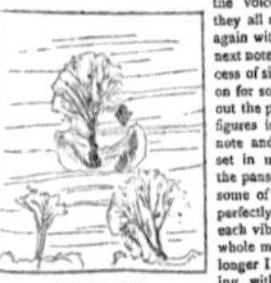

TREE FORMS.

BRITISH CAPITAL IN MEXICO.

THIS country has been of late investing a large amount of its spare capital in Mexico. Last year alone well on to fifteen millions sterling were sunk in railways, lands, mines, public securities, banks, and other enterprises, which with previous investments constitute, as Sir F. Denys says, in a report which he has just forwarded, "a tremendous mortgage on the resources of the Republic, rendering the maintenance of law and order a matter of almost as much importance to Great Britain as to Mexico itself." Specially over the railway system of the country we have got a firm hold. The Mexican Railway, the Interoceanic, and the Mexican Southern are English companies; the control of the National is in English hands; the Tehuantepec is being constructed with British capital, and the majority of the first mortgage bonds of the Central have passed into English hands. The money for the construction of new railways, concessions for which have been granted by the Government, will probably be sought for in London. The drainage of the valley of Mexico has been undertaken by English capitalists, English banks are about to be established, several million acres of land are owned by British subjects engaged in cattle raising, and a large proportion of the capital raised for mining enterprises has been found in London.

It is curious that while British capital is thus being lavishly invested in Mexico, American capital, except that interested in mines, is being gradually withdrawn. The reasons given by Sir F. Denys should make British investors a little cautious. So far as this withdrawal is due to the enormous and profitable field for enterprise open to Americans in their own country there is no occasion for alarm, but other causes are also at work. It is significant, for instance, that we should be reminded that "Mexico being close to the United States, the attractions it may offer are not so much appreciated because they can be readily seen and verified;" while, so far as the British investing public are concerned, "the distance is too great for people to come and judge for themselves of the soundness of the undertakings and ventures for which specious prospectuses are issued. The unscrupulous promoter, therefore, has an admirable field for his operations in Mexico." As a Mexican "boom" is shortly expected, and the British public will be invited to subscribe to all sorts of schemes, concessions for which have been freely granted by the Government, a word of caution does not come amiss. The problem, of course, is to discriminate between the sound and profitable schemes and the unsound and worthless ones; and, unfortunately, too many investors will only gain their information at the cost of their pocket.

Mexico is without question a land of large possibilities, and the Government is doing its utmost to develop its resources and maintain its credit abroad. Owing to the tranquillity which it has enjoyed for thirteen years and the opening up of the country by its railways, the revenue shows each year a satisfactory increase. In 1888-9 the receipts of the Federal Government were just under five millions sterling. The expenditure, however, is also expanding, and in the same year exceeded the revenue by £600,000. There is little chance of the accounts being balanced by a reduction of the expenditure, as the present figures are said to represent the lowest sum at which the government of the country can be carried on. The subvention to railways is very onerous. In the Budget for 1889-90 £462,000 is set apart for this purpose, and next year £697,000 will be required to meet the subsidies. The Government has undertaken liabilities to the various companies amounting in the aggregate to 33 millions sterling. In order to redeem the subvention certificates the Government is said to be negotiating for a new loan of £6,000,000, at 5 per cent. interest, and 1 per cent. sinking fund guaranteed by 15 per cent. of the Customs receipts, an operation which would effect a saving to the Government of 8 per cent. of the Customs receipts under pledge, besides giving them a large sum in cash. "The railway subventions," as Sir F. Denys says, "are the sword of Damocles suspended over Mexico's finances, and it is imperative for the Government to find a way to meet these liabilities." Were that achieved, there is every reason to regard the future of Mexico hopefully. Its external and internal trade is growing rapidly, mining is active, railways are being built, banking facilities are being extended, and generally the country is being developed, and "if proper caution and discrimination are exercised, there need be no fear on the part of the public abroad in embarking under the present Administration in those mining, agricultural, or financial enterprises which offer reasonable prospects of success."

The Ladies' Corner.

THE QUEEN'S BREAD.

The Queen's partiality for Viennese and French bread runs into all sorts of shapes. There are long French loaves, and twists and rolls, and the Viennese bread is shaped into all sorts of curves and twists. There is one roll made like a little mannikin. This is supplied for the edification and amusement of the Queen's small grandchildren when they sit at her Majesty's table. The Queen is always supplied with this bread when at Buckingham Palace. Her baker is Mr. S. Petrozywalski, a Polish refugee, in whom the Prince Consort took a great interest. This fancy bread is only supplied for the Queen's table; for the rest of the household the Palace baker bakes. Some of the larger loaves supplied to the Queen cost tenpence each. She did have this bread from London sent down to Windsor, but owing to the late arrival of the train that conveyed it, or something of the sort, this was given up. When the Empress Frederick was staying with the Queen, rye bread, of which the Empress is fond, was sent to the Palace. The Queen's bakers have always been able to satisfy

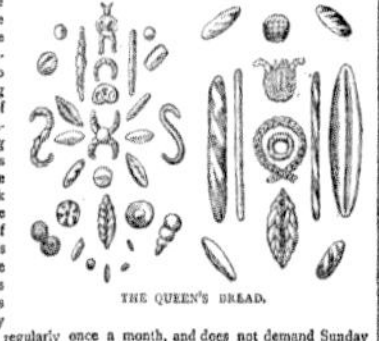

THE QUEEN'S BREAD.

the Queen. She pays regularly once a month, and does not demand Sunday bakings. When some of Mr. Petrozywalski's customers have grumbled that they didn't get bread on Sundays her Majesty's forbearance was quoted, and this usually stops their complaints. The same baker also supplies the Princess of Wales and other members of the Royal family, and though the Queen always has her confectionery and cakes made in her own kitchen the Princess of Wales and the Duchess of Connaught, who share the characteristic of having a sweet tooth, occasionally order lunch cakes and other good things. The Queen, it may be mentioned, doesn't like freshly-baked bread. It is always a little stale.

TO-DAY'S NEW BOOKS.

(Publishers would greatly oblige by stating the prices of books. Any notices under this head do not preclude further review should such seem to be necessary.)

CLIFFORD, the Rev. H. M., M.A. "Handbook to Exodus." (Henry Frowde.) [A volume of "Clifford's Old Testament Manuals," printed in parallel columns, the one containing the text, the other the Notes. Preface and Appendix. Cloth. Pp. iv. 128. Price 1s. 6d.]

DAVIS, JOHN FRANCIS (Editor). "George Sand: La Mare au Diable." (Librairie Hachette et Cie.) [Edited with an introduction, numerous notes, and a vocabulary. Cloth. Pp. viii., 166. Price 1s. 6d.]

KORTE, G. "Elements of French Commercial Correspondence." (Librairie Hachette et Cie.) [Part II., containing examples of circulars, letters of introduction, cheques, bills of exchange, &c., and a complete English-French vocabulary of commercial terms, comprising some 4,000 words. Cloth. Pp. 122. Price 1s. 6d.]

"THE METROPOLITAN YEAR BOOK, 1890." (Cassell and Co.) [London is here treated in its various aspects; the municipal and local, the commercial, the imperial, the ecclesiastical and educational, and the social and miscellaneous. Cloth. Pp. xli., 320.]

THE "TIMES" AND THE GOVERNMENT.

LAST NIGHT IN THE COMMONS. (BY AN OBSERVER.)

WITHOUT any of that gentle proem by graceful though not eloquent gentlemen in elegant uniforms with which the abrupt transition from the dolce far niente of the recess to the hard work of the session is usually softened, we rushed last night straight into the middle of the most exciting topic of the day. The House has rarely been in so pugnacious a spirit on the first day of the session. The vigorous whips on both sides had resulted in an assemblage of members which would not have shamed a great Government night in mid-session. By two o'clock every bench was occupied with its hat, and when the Speaker came in to prayers a great throng followed close on his heels. The usual ceremonial followed. Black Rod was seen approaching, and was without hesitation most incontinently shut out. Undismayed, Black Rod knocked, and on his intimation that their presence was required in the Lords, the Commons obediently adjourned thither, and listened to the Queen's Speech. The House then adjourned until four o'clock.

At that hour members assembled once more in still greater force. The front Opposition bench ill-sufficed for its contents; and there was plainly something in the wind. After the usual formalities of the first day of the session—signing the paper of motions and re-enacting the standing orders—the new members were called to the table by Mr. Speaker. The four new Liberal members—Captain Verney, Mr. Seymour Keay, Mr. Lea, and Mr. Morton—were greeted with vociferous cheers by the crowded Opposition benches, and the Tories gave tongue as loudly over their one escaped lamb—Mr. Loder from Brighton. Mr. Loder, however, was eked out by Mr. Chaplin, who came forward with a timid modesty which would have befitted the newest of new members, encouraged by shouts of "Go on! Go on!" from the delighted House.

After a few skirmishing questions we passed into the dreary land of notice of bills and motions, and it was close upon six o'clock—after a warning had been moved for the election in Glamorganshire—when Sir William Harcourt rose to move that the House regarded the publication of Mr. Parnell's letter on April 18, 1887, and the comments thereon, a breach of privilege. The House had been emptying but immediately re-filled. Sir William Harcourt was most intensely serious. No a joke—scarcely a jeer—dropped from his lips during the whole of not very long speech. He began by explaining why there had been a delay in moving this motion—intimating that the Speaker had asked him for some such an explanation—and immediately gave the cue to the Opposition by laying the blame for the delay on the Government. This, as we shall see, was the line adopted throughout the evening on the Opposition benches. "We, the Opposition, always clamoured for an early investigation of the charges, but the Government always threw every possible obstacle in our way. No motion of the kind could be moved until the letters were legally pronounced to be forgeries. This did not happen until February 4. We are therefore acting as speedily as possible." The delay was no obstacle: there were precedents to show that such motions could be moved in regard to events not only of previous Sessions but of previous Parliaments. (Sir John Gorst afterwards bitterly complained that Sir William quoted none: and the reason why was not obvious.) He then went on to show that the libel was essentially a breach of privilege, because it aimed at influencing the opinion of the House, and read from poor Macdonald's answer at the Commission and the Times leader of April 18, on the night of the second reading of the Crimes Act, to show that the object was the passing of that measure. He then quoted many precedents of cases which had been recognised as breaches of privilege—Justice Lopes' "disreputable Irish band," Plimsoll's placards, and Bright's accusation of alliance between Tories and "Irish rebels,"—and argued that if there were adequate and necessary occasions, then a fortiori was this: "What is called an apology" had been given in court; but "in this place there must be some reparation." Sir William ended by appealing to the Act of Union for equality of treatment to the Irish, and regretting that the Leader of the House had not performed the duty in his place.

The disappointment of the House was immense, when, in spite of loud cries for "Smith! Smith!" Sir John Gorst was set up to reply. He had three arguments for moving his amendment, which was a virtual negative. 1. The lapse of time. No precedent existed for a delay of three years. 2. The libel had been a subject of legal process, and was "still to a certain extent sub judice" (words of which Mr. Parnell took a note). They must wait for the Report. (At this instance a mouse strolled up the floor of the House in a leisurely way, and the consequent laughter much puzzled poor Sir John.) 1. The case was not bad enough. His hon. friend the Chief Secretary was accused of innumerable crimes weekly, but Parliament took notice. Mr. Parnell had the ordinary courts. "These forged letters," played out. Hon. members have made all the political capital out of them that they can."

After a short, manly speech from Mr. Reid, Mr. Gladstone rose shortly after seven to answer Sir John Gorst. He took his arguments in reverse order: 1. This was not a usual charge. The usual charges are constructive. This is definite and direct, that meant absolute political death for Mr. Parnell. 2. His second argument clashed with his first. It could not be both too soon and too late. As for the "too late" argument "the maker of delay are her Majesty's Government." After an eloquent vindication the "solidity and eminence" of the "uncrowned King," Mr. Gladstone touched upon the action of the Times, which "came very near to absolute falsehood." A great debt was unpaid—a debt to Ireland, to England, and to the House of Commons.

Mr. Balfour then considerably improved upon the case of Sir John Gorst. His chief point was that the Times, by paying 40s. into court, admitted the forgery of the letters. Why was not this motion moved then—three months ago? He then defended the Times, condemned the tendency to minimize the other charges in the Report, and enlarged on Sir John Gorst's argument that the Radical press was far more prone to false accusation than any other.

Mr. Labouchere then treated us to a few more "Pigottiana," which he dealt out with considerable skill. He had traced, he asserted, to the Times office a number of bank-notes sent to Dublin after Pigott's exposure. Coupling this with the telegram from Madrid, he said that he believed the Times guilty of trying to get Pigott to Madrid and to support him there. After a forcible contrast between the treatment of Mr. Walter and Mr. Parke, he ended by saying that he respected friendships, and would not ask Mr. Smith to speak. (Mr. Smith remained silent.)

The debate was continued by several less important speakers, and Mr. Clarke closed the case for the Government with a remarkably effective speech, pleading the inefficacy of the law of privilege and the want of any proper procedure. He made an honest acknowledgment that Mr. Parnell had wholly cleared his character, and asked what they proposed to do with the Times if they passed the motion?

Mr. Parnell then rose, and in a clear and simple speech, a quarter of an hour's duration, touched the House to the quick, especially with the pathos in which he dwelt upon the long years of agony and anxiety taken in clearing himself from a charge "still regarded" (he said bitterly) "as sub judice." As the charge of being too late, had they not pressed for a Special Committee, a could they ask for a privilege motion until the forgeries were proved? The forgeries could not be proved until they found the forger; and the Times refused to reveal him in the ordinary courts. Had the Government granted the Committee, the whole thing would have been ended in forty-eight hours. "If I had gone before a jury I should have been left as I was with all the experts against me. I think was wise." The Committee was refused, he said, fiercely, because they wanted the forged letters as capital. "You did not care whether they were forged or not." And then a large "fishing" inquiry was secured. "The letters were forged to obtain it." "Leader as I am of a minority, I should be sorry to treat my most powerful opponents with the incredible meanness and cowardice."—(lost in shouts.) "I am sorry for you." He ended with demanding that the epithet "forged" should be inserted in the Government's amendment.

Then a surprise occurs. Mr. Smith rises with the greatest affability, and in his most gently paternal manner, accepts the amendment of Mr. Parnell, and expresses on behalf of himself and all his friends their detestation of the forgery and their satisfaction at Mr. Parnell's clearance. It is generally agreed that Mr. Smith's action is the only bright spot in a very dark even for the Government. On a division, the Government amendment was carried by 260 votes to 212—a majority of only 48 for the Government in a most essential issue. So this stirring evening closes with a cry of triumph from the Opposition.

"To me the whole matter becomes more wonderful the longer I study it," said Mrs. Hughes, looking with thoughtful eyes on the strange beautiful forms. "There lies a whole hidden world behind these forms, which the future may perhaps reveal. When I walked in the garden on one occasion after I had been singing a number of daisy forms into shape, and the petals had become so familiar, I had the feeling as if I had only to sing under the stalk of one which was yet in the bud, and it would open like the flowers produced by the voice."

An excerpt from "Voice Figures and What They Are: An Interview with Mrs. Watts Hughes," *Pall Mall Gazette*, February 12, 1890, as quoted by H. S. Olcott in "Mrs. Watts Hughes' Sound-Pictures," *The Theosophist*, September 1890

The Theosophical Society was founded in 1875 by Helena Blavatsky, its principal thinker and architect, and Henry Steel Olcott, who served as the society's president until his death in 1907. The Theosophist, edited by Blavatsky and Olcott, began publication in India in 1879 and served as the Theosophical Society's official journal. Blavatsky would have still been an editor when the following article was published; she died in 1891.

THE THEOSOPHIST.

VOL. XI. No. 132.—SEPTEMBER 1890.

THERE IS NO RELIGION HIGHER THAN TRUTH.

[*Family motto of the Maharajahs of Benares*]

MRS. WATTS HUGHES' SOUND-PICTURES.

H. S. OLCOTT.

AN acoustical phenomenon has been stumbled upon by a Welsh lady of position—Mrs. Watts Hughes—which is provoking the interest of men of science and the wonder of many of the profane. It is, in truth, very striking: most interesting to the student of occultism. To him, it is another proof that we are living in a time of, what might be called, akasic *osmosis*—the infiltration of the astral into the physical plane of consciousness. The partition between the two worlds is proving itself as porous and penetrable as the membrane between two fluids of different densities in the familiar chemical experiment. The deeper potentialities of thought, life-force, sound, chromatics, heat and electricity, are being successively revealed. To cap all, there is Keeley's inter-etheric force, whose titanic energies are just being demonstrated. On the plane of physics, on that of mental dynamics, that of psychical potentialities, and on the vastest of all—the limitless, in fact—that of spirit, the evidence of this cosmic action forces itself upon one's attention and strikes the thoughtful observers with awe. Each day brings its surprise, each month its contribution to the sum of human knowledge. Yesterday it was the phonograph, the telephone and the electric light, today it is the startling discoveries in thought-transference and other phases of psychical science, the transmutation of sound into form and of color into sound. What it shall be tomorrow, who can guess? The one thing to rejoice the occultist is that each forward step which science takes brings man nearer the threshold of that "Borderland" beyond whose shadows the sun of spiritual truth is ever shining. To him, such phenomena as those of mediumship, hypnotism, clairvoyance, thought-transference, and those of the quasi-physical class of which Mrs. Watts Hughes' are an example, have a moral value beyond compute,—one which eclipses their every other, viz., their evidential bearing upon the problem of the higher self and the higher life. This action and progress, this increased knowledge of Nature's arcana, this wealth of discoveries, are bad for theologies. As the inter-relation of forces and the possible osmosis of planes of being become more and more evident, the unity of nature, of the human race, and the superiority of truth over dogma, tend to sweep away all that divisions of creeds, castes, races and human interests have done to obstruct the inflow of divine truth and the recognition of human brotherhood. For those who wait for such a consummation, the Buchanans, Dentons, Foxes, Didiers, Edisons, Charcots, Bernheims, Hares, Zöllners, Crookeses, Tyndalls, Keeleys and Hugheses are true benefactors of mankind. The pictures drawn out of the world of invisible forms by the vibrations of Mrs. Watts Hughes' glorious voice, are the subject of our present comment and will now be described.

Our narrative is compiled from the account which appeared in the *Pall Mall Gazette* of February 6, and the illustrations are from electrotypes of the original cuts in that journal, kindly furnished us by the Manager, at the instance of my friend Mr. Stead, to whom I applied.

The history of the discovery, as given by Mrs. Watts Hughes herself, proves that it was, like that of the principle of the spectroscope by Frauenhofer, and that of the phonograph by Edison, "accidental," in a way. She had been endowed by nature with a superb voice and had always pondered deeply upon the laws of acoustics, yet the present discovery had not been previously thought out, but stumbled upon. She had read about "Chladni's figures," and was experimenting with her voice upon sand sprinkled over a plate of glass, to see what effect the vibratory tones would have upon it, when—as she tells it—

I saw one day with intense surprise that the grains of sand with which I experimented formed themselves into a geometrical figure not unlike those which Chladni discovered. In fact, the figures which I then produced were Chladni's figures discovered over again. I continued my investigations, and slowly and gradually discovered that by singing certain notes into the eidophone over the mouth of which the disc is placed, I can sing various substances, such as sand, lycopodium, or coloured liquids into certain figures. Every single note produces a figure, in which the vibrations of the voice are recorded by clear and regular lines. According to the pitch, intensity, and the duration of a note, however, the form of the voice-figures differs.

The *Pall Mall* interviewer then asked her whether a certain figure is always produced by singing a corresponding note. The following conversation then ensued: —

"Then, is a certain figure always produced by singing a corresponding note?" "The daisy and Pansy and all the geometrical figures can be produced by any note of the scale; but there are many other points to be considered in the mode of production, which it would take too long to explain. Now I will show you how I work." Mrs. Hughes sat down in front of the eidophone, on which a small quantity of fine powder had been scattered. A deep, full note was sung into the tube, and immediately a miniature storm raged on the disc. Tiny clouds of dust arose, rolling and whirling about as when a hurricane sweeps over a dusty high-road. Slowly the chaos was reduced to order, and when the last vibration ceased, an accurate, clear geometrical figure lay before us, formed of the yellow powder on the dark disc.

"Now I'll change it into another figure," said Mrs. Hughes, and once again the storm began as a rich, melodious note went up through the tube. The forms produced in sand or powder are, of course, unstable, and change or vanish when the instrument is moved, but in moist colour they can be perpetuated, and the experiments with paste are, therefore, infinitely more interesting. Thus, for instance, when a daisy is to be created, the substance placed on the disc creeps together in the centre of the membrane at the command of the first note, which is obeyed as unhesitatingly as the bugle horn in the soldier's camp. Another note follows of a different calibre, and out from the centre all round shoot small petals shaped exactly like those of the daisy. But perhaps the note has not been quite so full as it should have been, and the delicate petals are not quite as symmetrical as they ought to be. Then comes one of the most marvellous parts of the process, for once more the voice commands, and immediately they all rush back and amalgamate once again with the centre, to reappear at the next note, smooth and shapely. The process of singing waves in and waves out goes on for some time, till the last wave brings out the perfect daisy form. The size of the figures is determined by the pitch of the note and the quantity of substance to be set in motion; and as with the daisy and the pansy, so it is with all the other figures, some of which are marked with the most perfectly shaded fluted lines, representing each vibration of the voice. "To me, the whole matter becomes more wonderful the longer I study it," said Mrs. Hughes, looking with thoughtful eyes on the strange beautiful forms. "There lies a whole hidden world behind these forms, which the future may perhaps reveal. When I walked in the garden on one occasion after I had been singing a number of daisy forms into shape, and the petals had become so familiar, I had the feeling as if I had only to sing under the stalk of one which was yet in the bud, and it would open like the flowers produced by the voice."

"In the charming drawing-room below we loitered once again over the screens and shades and tiles on to which the strange figures had been sung, and about all of which there is a strange look of life and movement. Two charming little lanterns were set apart to be taken to Sir Frederick Leighton, who takes the deepest interest in the discovery; and as we went away it seems as if there were music even in the roar of the railway-train and in the creaking of the carriage-wheels in the street."

The lady is right in saying that a whole hidden world lies behind these forms. If she were clairvoyant she might have seen

PAGES 40–41: *Voice Figures, ca. 1885–1904, paint on glass with stamped metal frames, 1⅞ × 1½ in. (4.8 × 3.8 cm) each*

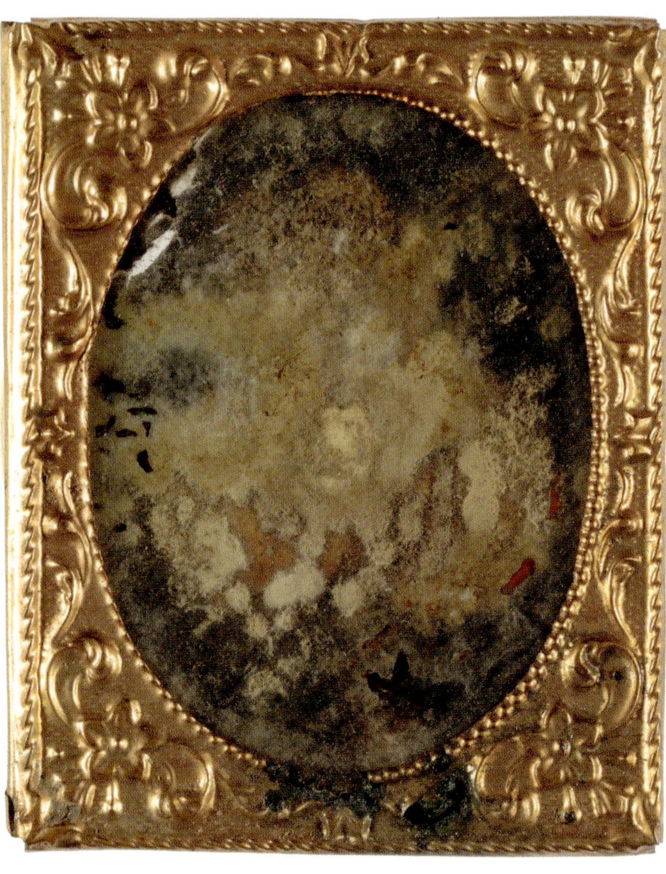

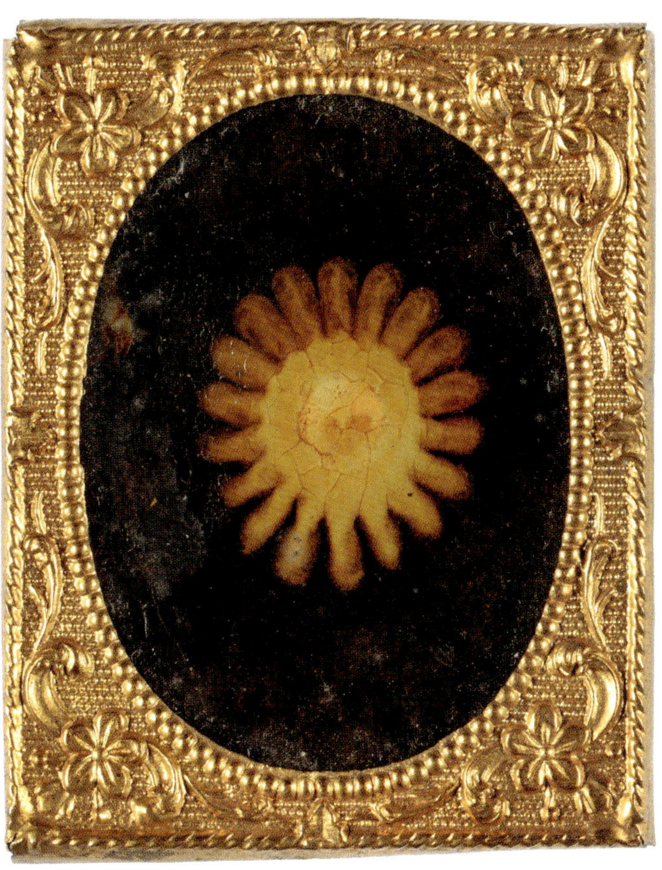

it. Many seers and seeresses have described it; none more intelligibly than Mrs. Denton, the queen of psychometers. By her exceptional introspective power she has been able to trace out in the galleries of the spatial ether the reflected images of past ages of our earth. She and her husband, Professor William Denton, call this side of nature the Great Psychometric Realm, and the latter with the enthusiasm of a prophetic scientist, details the inevitable lines of our future research:

"What a realm!" he writes: "the heavens of all ages for the astronomer; all the past of our planet, and its myriad life-forms for the geologist; all the facts of man's existence for the historian; all plants, from the fucoids, that spread their arms on the tepid seas of the primeval world, to the soaring cedars of California, for the botanist; for the artist, all the giant mountains that the rains of æons have washed away; . . . the smoking mountains, the flaming craters, and the spouting geysers, not of this planet alone" for the ever-growing soul to study and come to know (*The Soul of Things*, vol. 3, pp. 355–56).

Mrs. Denton identifies this realm as a plane of matter, yet far more sublimed than that of our plane of waking consciousness. She says:

"What a difference between that which we recognize as matter here and that which seems like matter there! In the one, the elements are so coarse and so angular, I wonder we can endure it all, much more that we can desire to continue our present relations to it: in the other, all the elements are so refined, they are so free from those great, rough angularities which characterize the elements here, that I can but regard *that* as by so much more than this the real existence.

"Something appears to me to be continually passing from our earth, and from all existences on its surface, only to take on there the selfsame form as that from which it emanated here; as if every moment as it passed had borne with it in eternal fixedness, not the *record merely* of our thoughts and deeds, but the actual, imperishable being, quick with pulsing life, thinking the thought, and performing the deed; instead of passing away into utter nothingness, that which is *here and now* forever continuing and eternized *there and then*" (*Soul of Things*, p. 346).

Psychometry gives us thus an apprehension of the registrative action of nature, by which she records in the ultra-material plane the minutest incidents of her evolutionary phenomena. We see here objective forms vanishing into latent permanency: reverse the process, and we can apprehend the objectivation of latent realities of form, color and sound. It is hard—for me, at least—to account for Mrs. Watts Hughes' voice-pictures on the purely materialistic theory of acoustic vibrations. Chladni's figures (fully illustrated in Prof. Tyndall's work on "Sound," p. 143) were obtained by setting up vibration in glass and metal plates, sprinkled over with dry, fine sand and lycopodium seeds; and with the same mixed with thin gum-water. The plates were fixed at one side in a clamp, and free at the other parts of their edges. In the case of the dry sand and powder, the patterns into which the particles arranged themselves under the vibrations of the plates were geometrical shapes, or combinations of wave-lines: when dampered, they sometimes took on the character of mossy growths. But to read the graphic description of the *Pall Mall* reporter and glance at the illustrative cuts, there seems to be some obscure formative agency at work, not quite identical with physical vibration. It is almost as if the harmonies of the singer's noble voice had, like a spell, evoked from a fairy-world the beauteous models from which natural evolution draws its supply of visualized shapes. What elfin bugle is this that has been brought out for our inspection; on what flowery bank of that inner world grew the daisies and pansies that have been photographed for our delectation? And this picture of what might almost seem a bit of arctic ice-scenery, with the towering, half-melted iceberg and the floating hummocks about it, how real it seems! [opposite] Readers of "Isis Unveiled" will recollect Prof. Tyndall's alleged description of the marvels he saw, of swimming fish and nests of cones, when he passed a beam of polarized light through the vapours of nitrates in a glass tube. He, too, if the account be true, must have had a peep into this hidden realm of ultra-substance. If I had never seen the creative powers of Madame Blavatsky shown in a number of cases where she brought out of latency and fixed in visible form images that existed in the astral light, I might not be so inclined to connect Mrs. Watts Hughes' voice-pictures with some psychical process going on within her brain. That she herself probably does not suspect this relation to exist, weighs as nothing against the theory: recent hypnotic experiments prove that the "two selves"in a person may be engaged in opposite and mutually independent acts of consciousness. Has she ever previously thought of a new voice-form which she subsequently evolved by her singing? And do contralto, tenor and bass voices create a different class of forms through the eidophone? It will be interesting to watch this estimable lady's future discoveries in this same direction of acoustical research. She deserves success.

Voice Figure, ca. 1885–1904, paint on glass, 6⅛ in. (15.5 cm) diameter

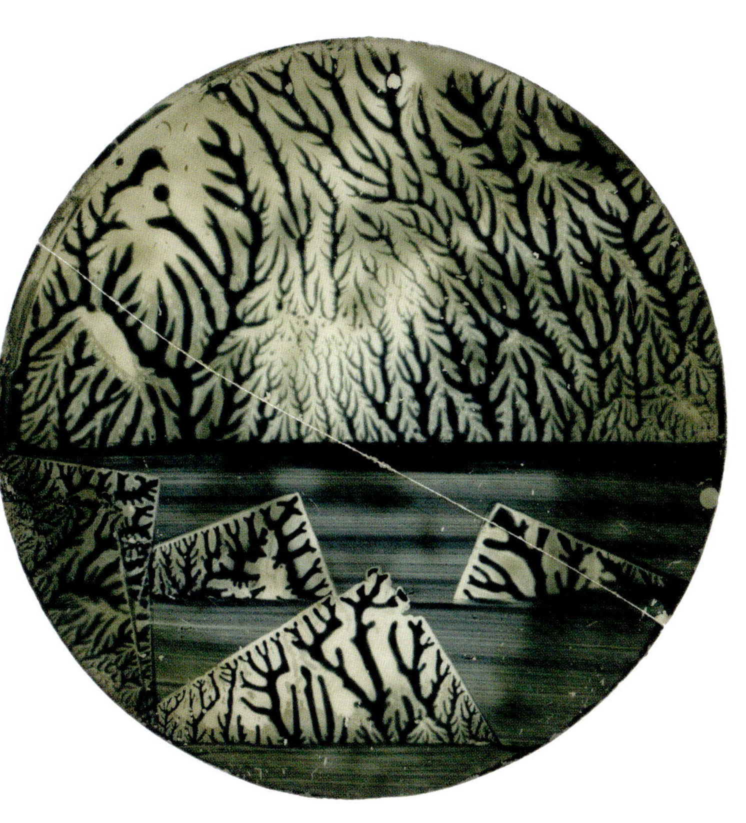

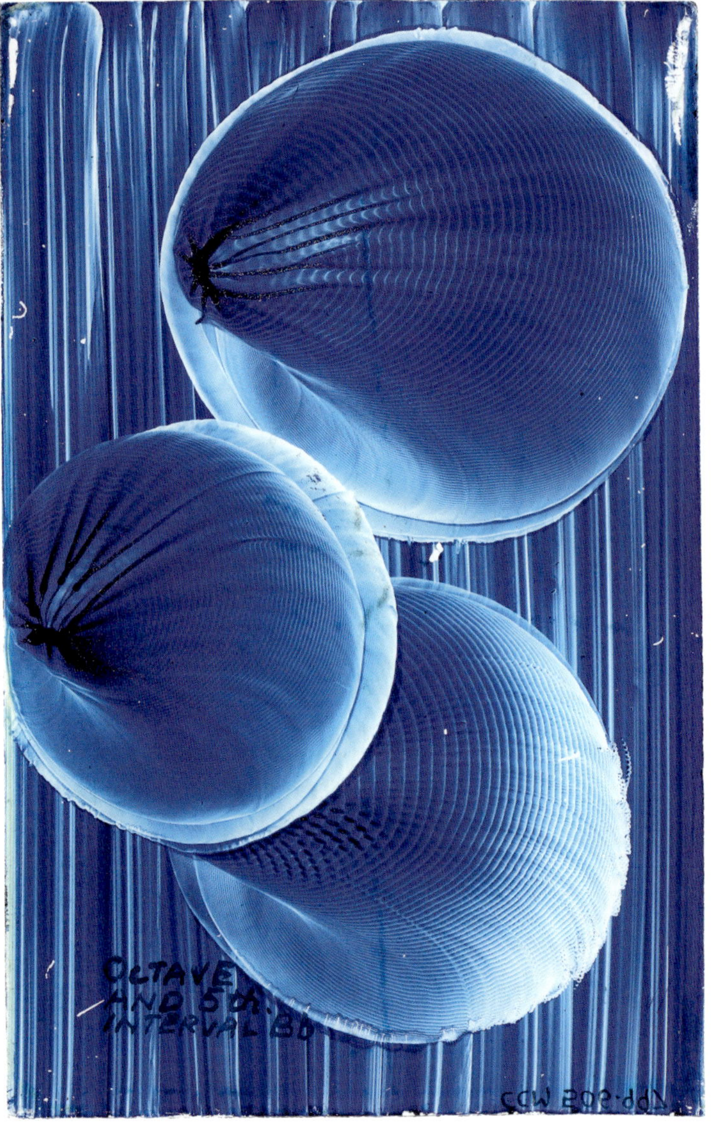
OCTAVE
AND 5th.
INTERVAL Bb
CCW 202·ddA

"We now pass to a large class of voice-figures quite distinct from all that we have so far noticed; viz. figures obtained upon plates of glass or other smooth surfaces brought into contact with the vibrating disk, both plate and disk having been coated with liquid color.

"In this class of figures I have been much gratified by obtaining very delicate yet perfectly clear wave-line impressions that constitute an actual and permanent record of every individual vibration caused by the voice, each tiny undulation of the surface of the disk standing registered with strict accuracy and with a beautiful minuteness that rivals the lines of a well-executed engraving."

—Margaret Watts Hughes, "Visible Sound,"
The Century Magazine, May 1891

VOICE FIGURE GLASS PLATES
1885–1904

Selections from the Watts Hughes collection at Cyfarthfa Castle Museum and Art Gallery, Merthyr Tydfil, Wales, UK. These Voice Figures were created by Margaret Watts Hughes using a hand eidophone and paint on glass. Dimensions appear on pages 86–87.

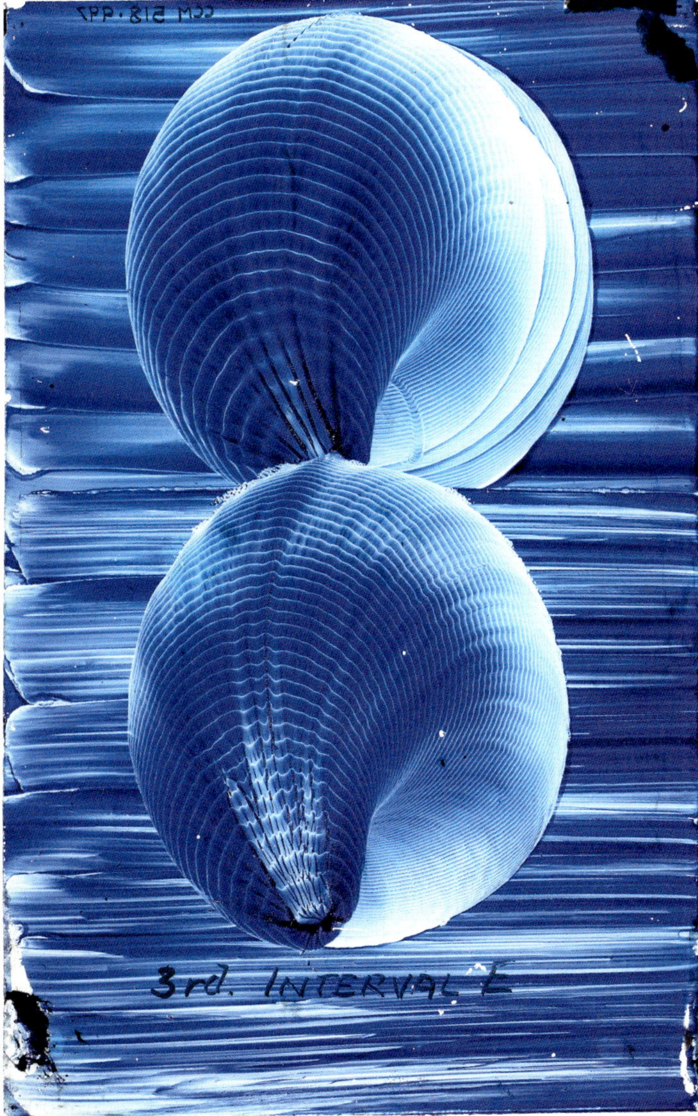
3rd. INTERVAL E

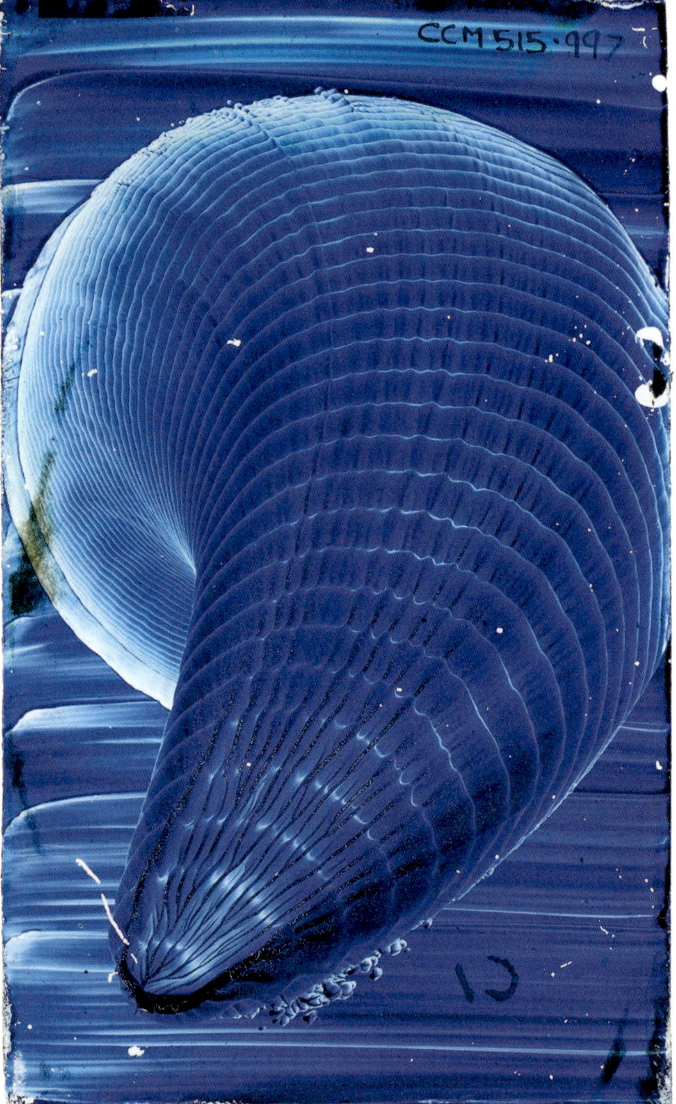
CCM 515·997

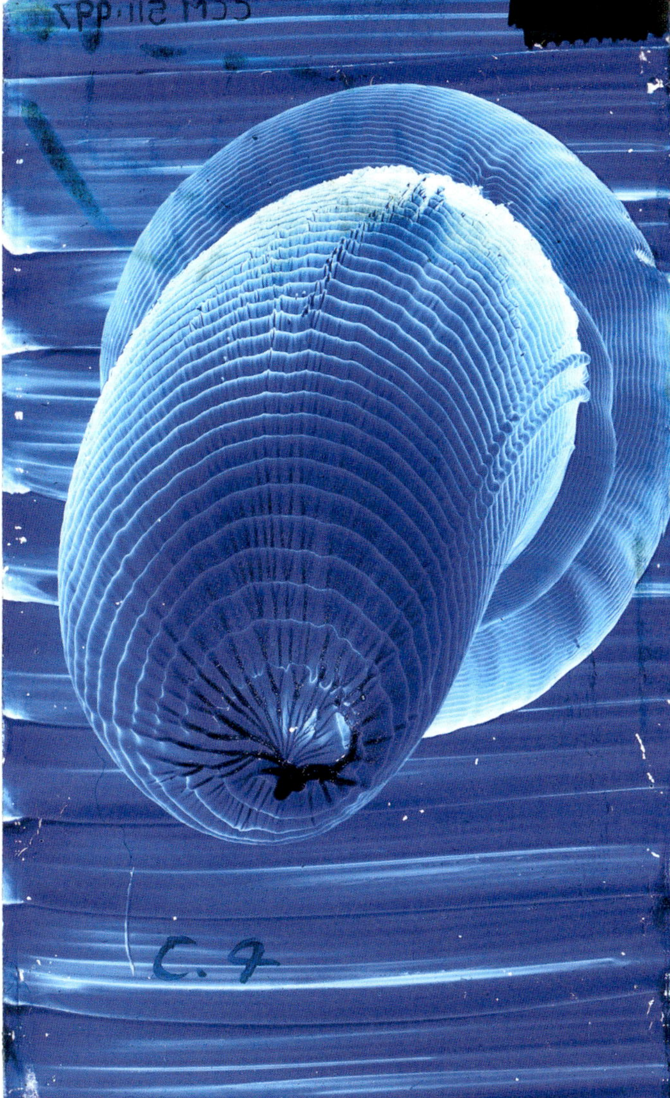
CCM 511-447
C. J

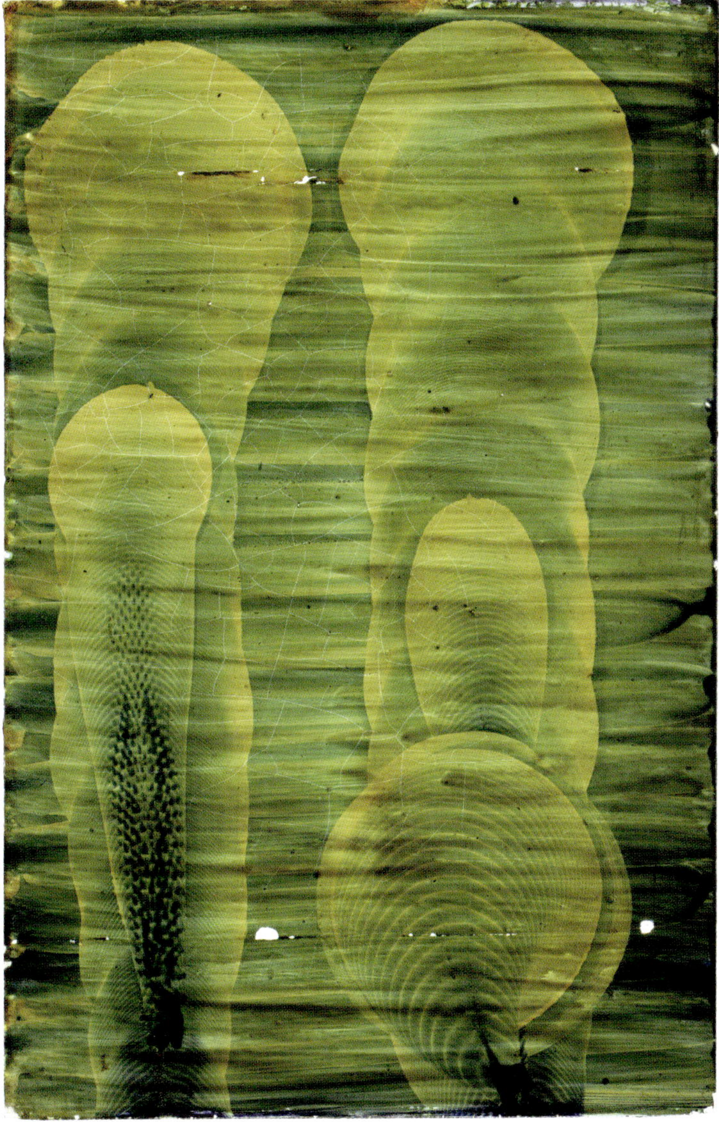

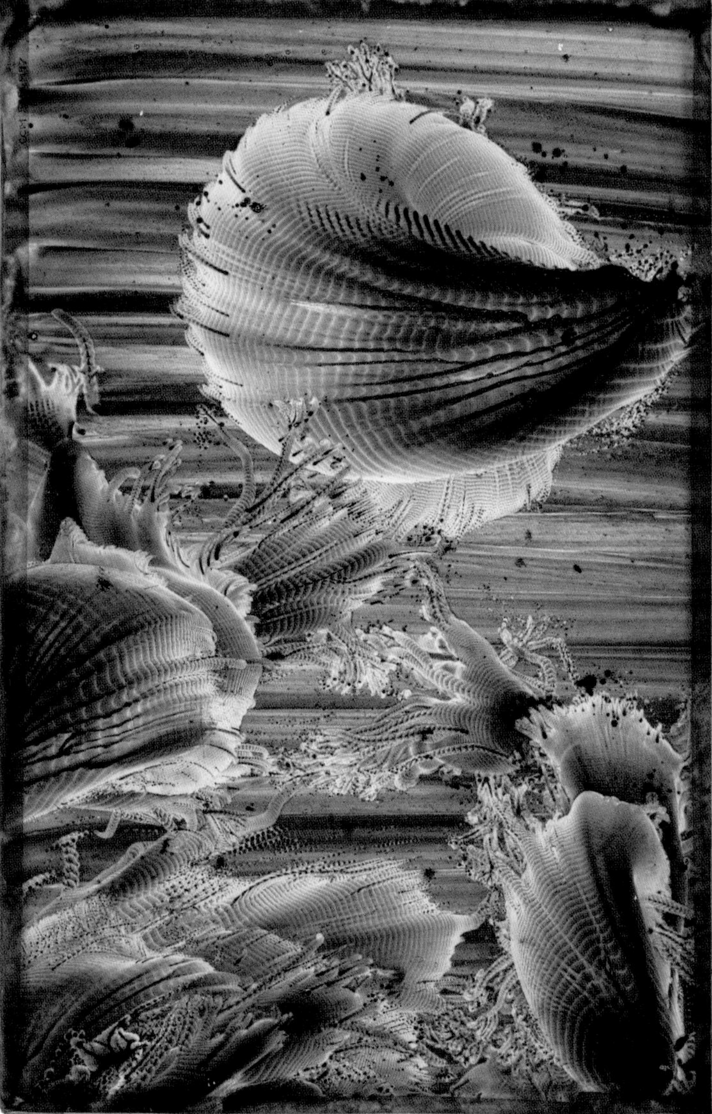

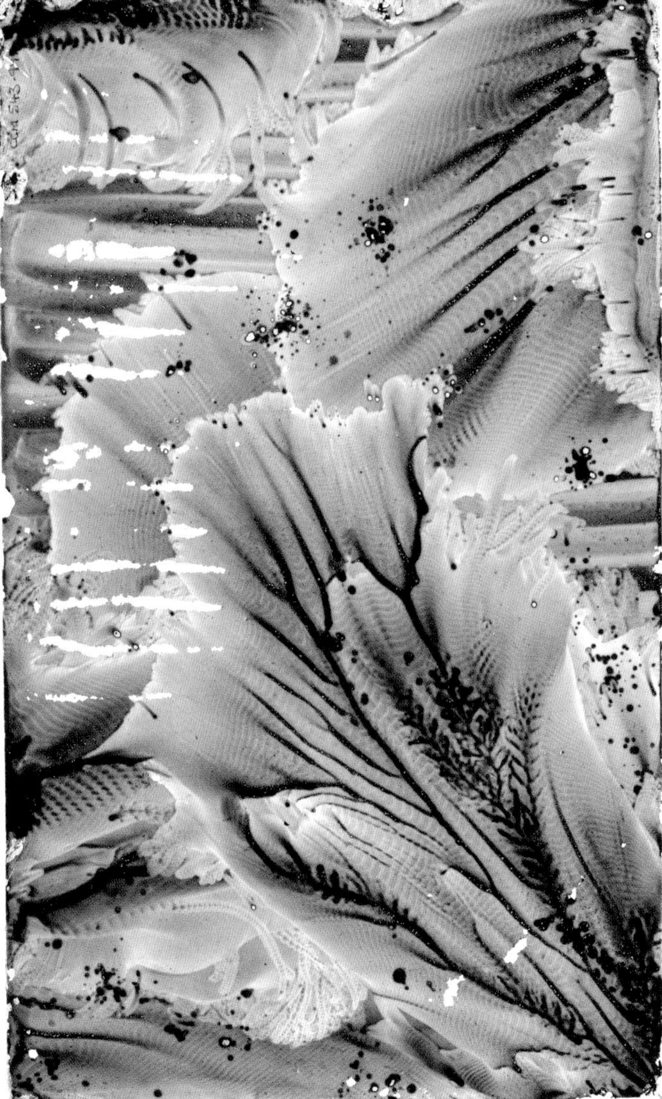

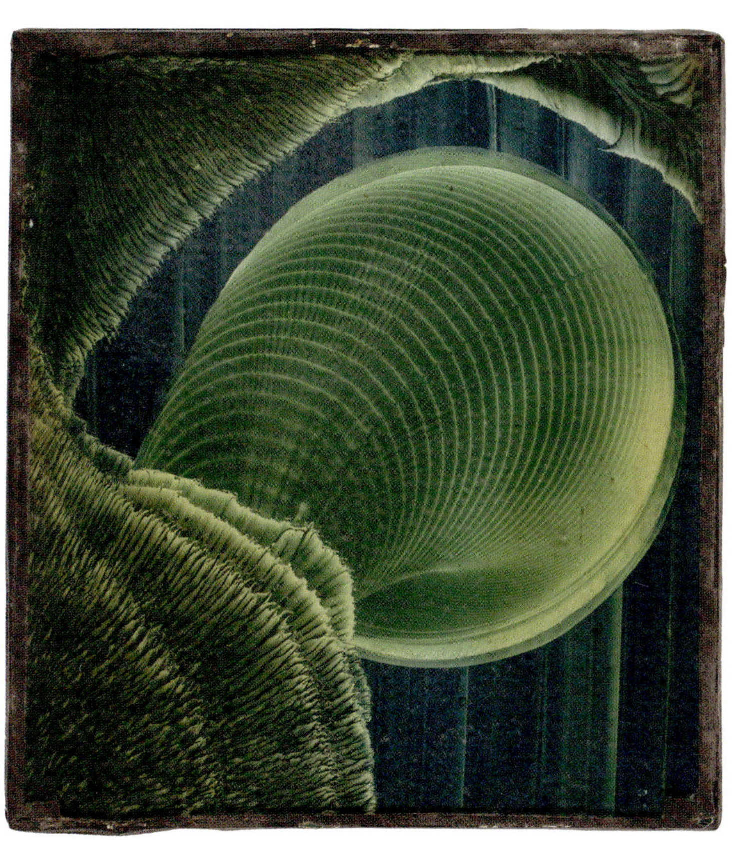

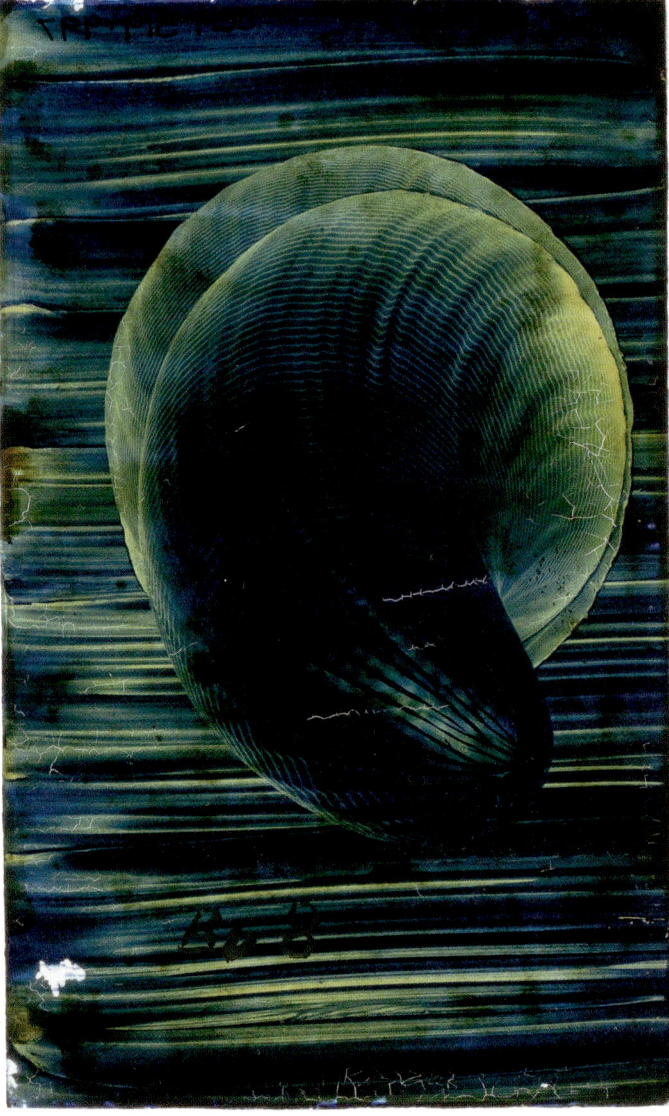

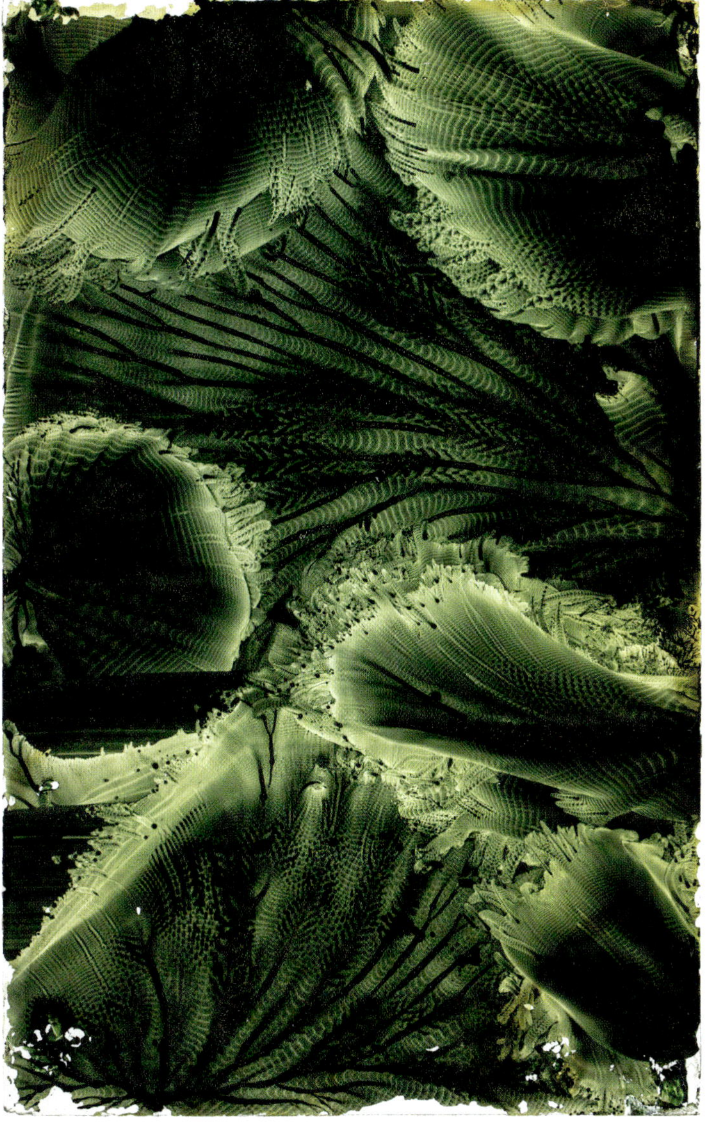

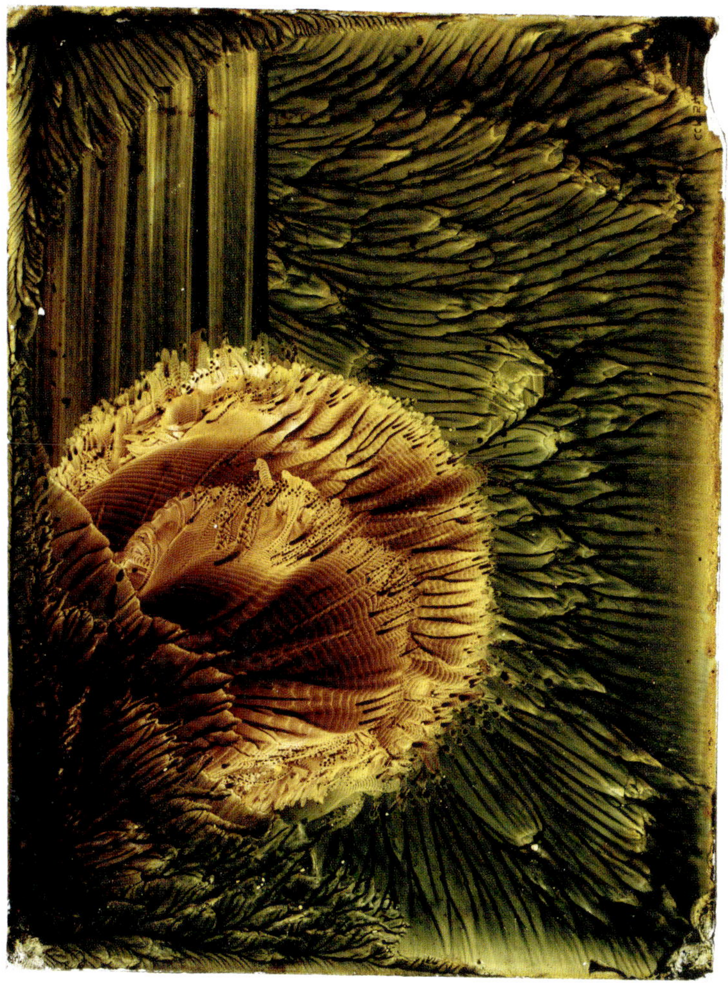

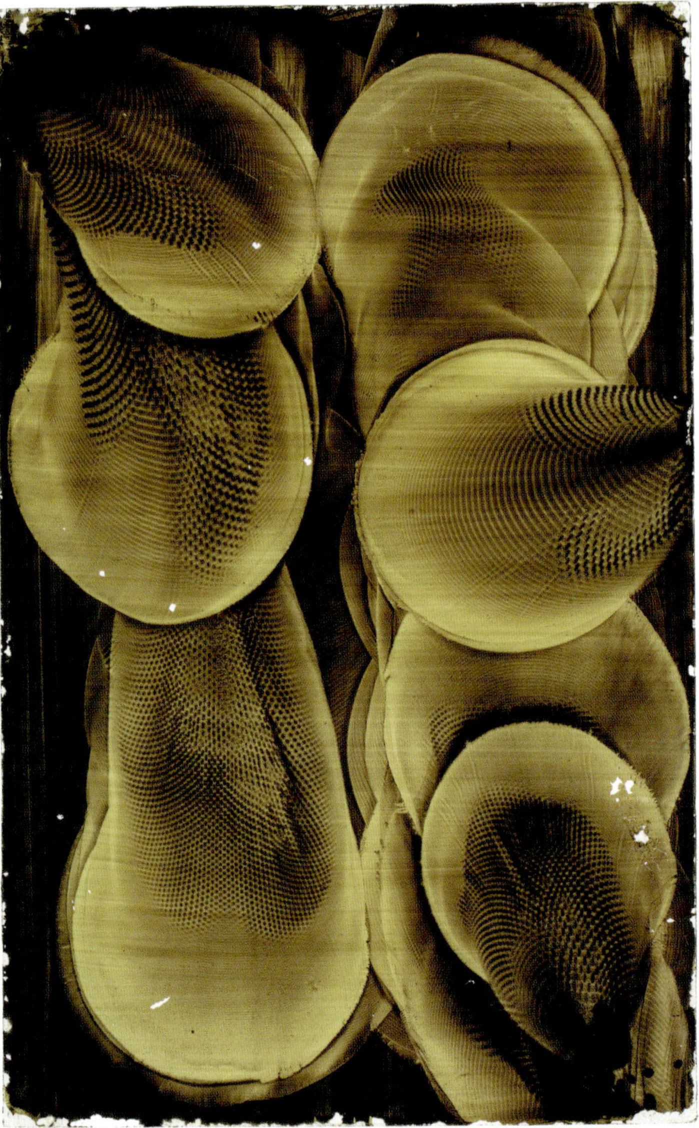

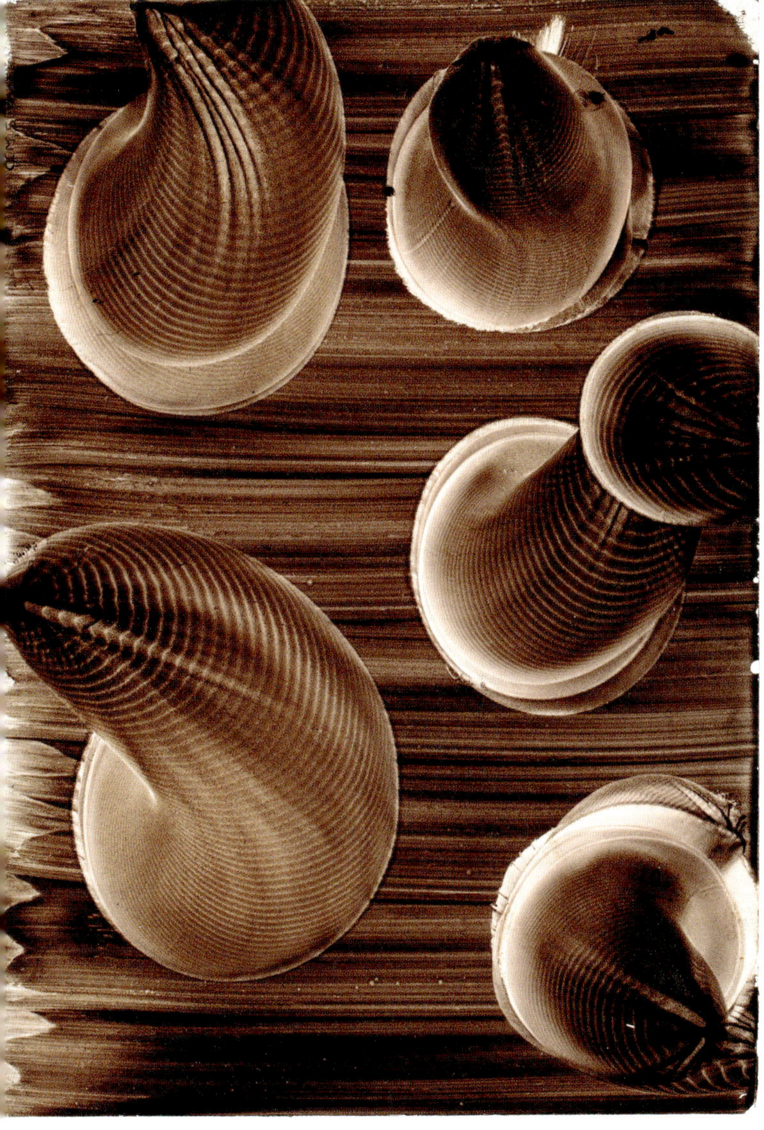

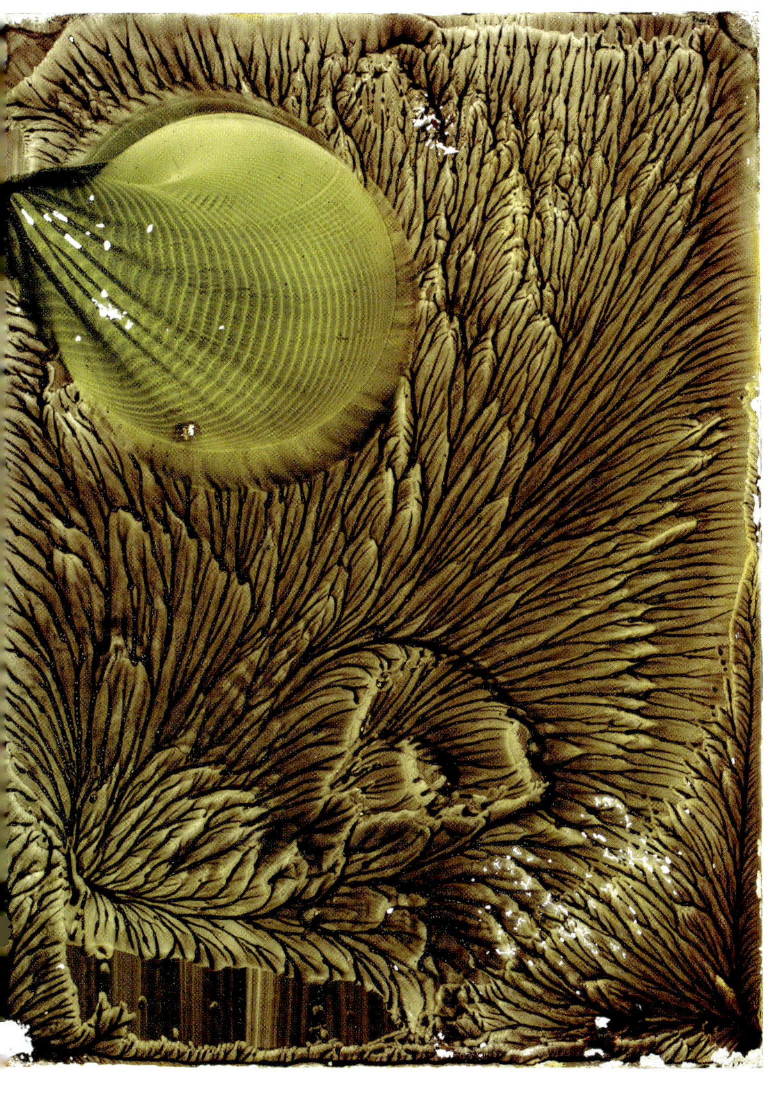

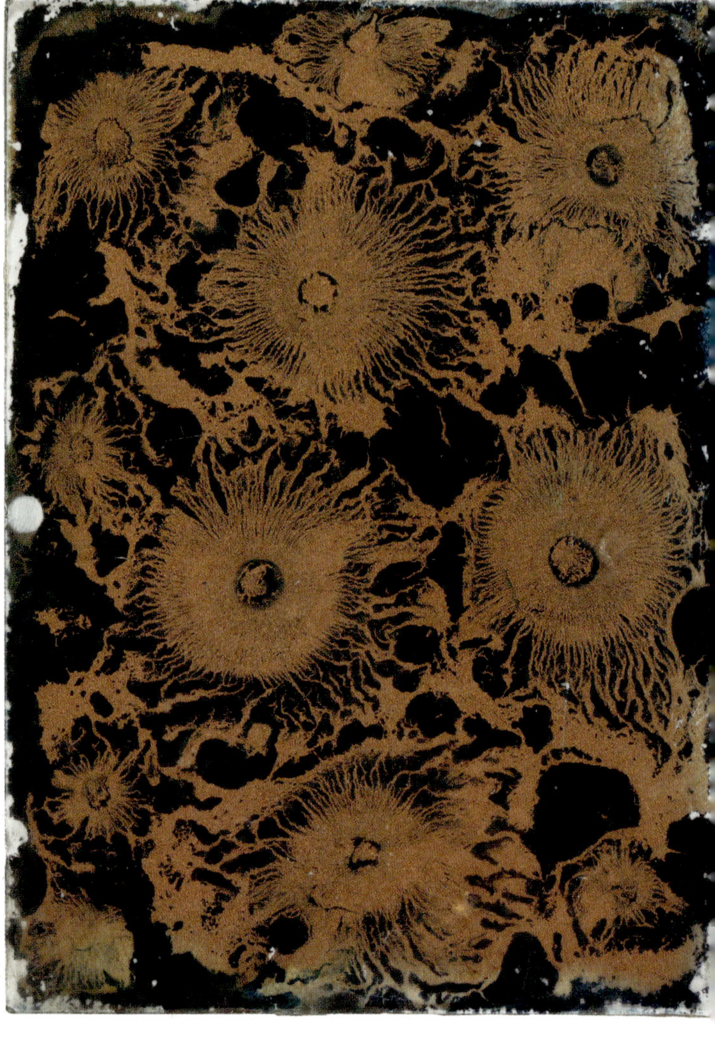

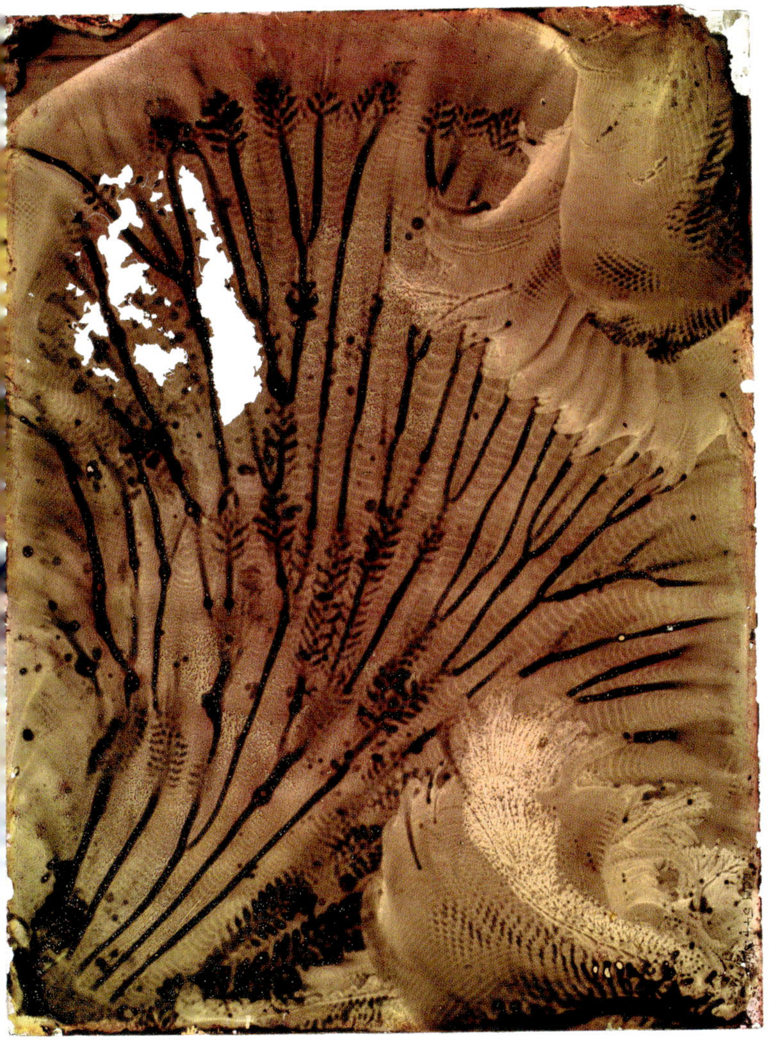

John Watts, the brother of Margaret Watts Hughes, wrote an unpublished biography of his sister and spent many years promoting her work. He used the following glass lantern slides to illustrate his lectures. The slides measure 3¼ × 3¼ in. (8.2 × 8.2 cm) each and are all in the collection of the Cyfarthfa Castle Museum and Art Gallery.

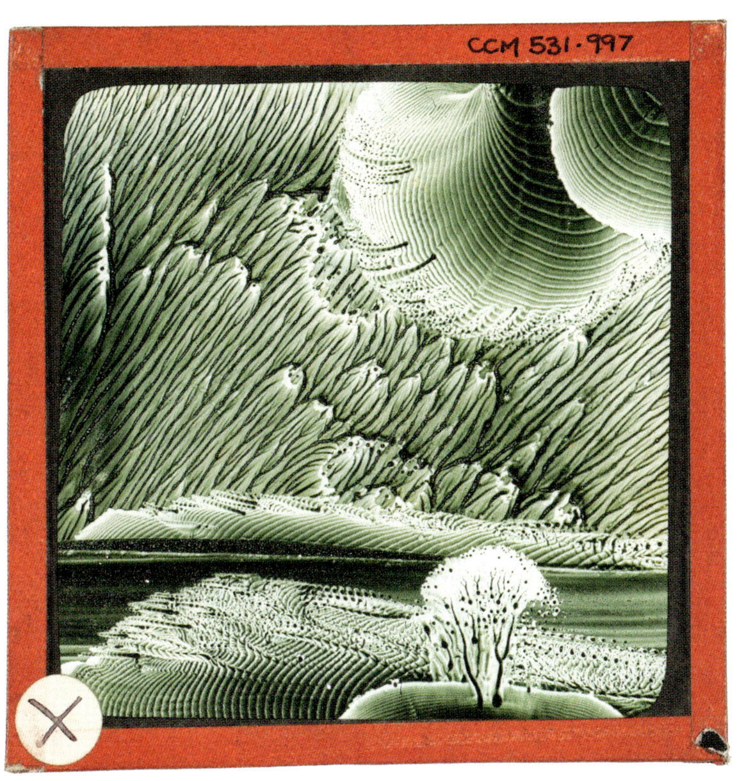

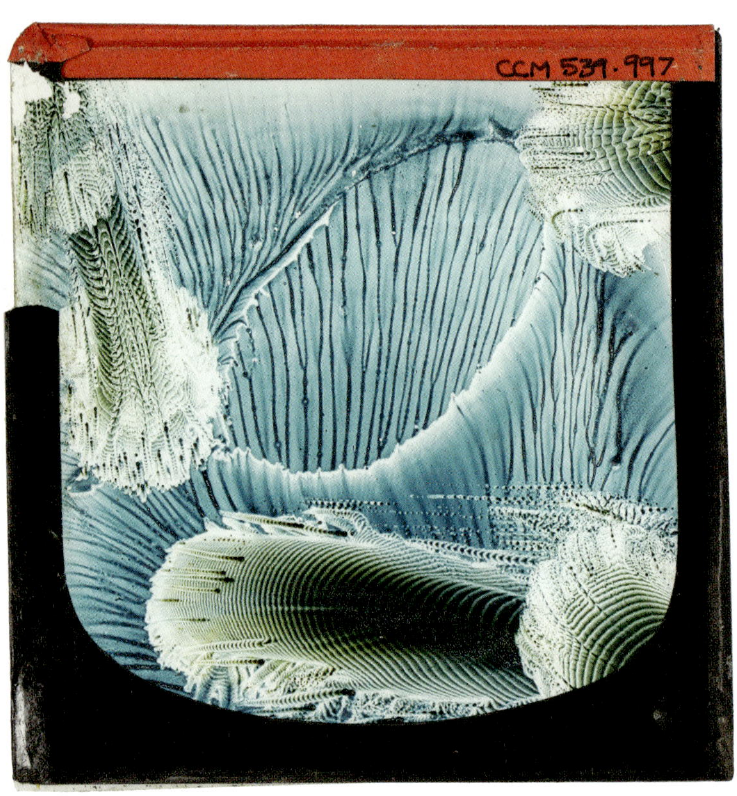

CCM 539·997

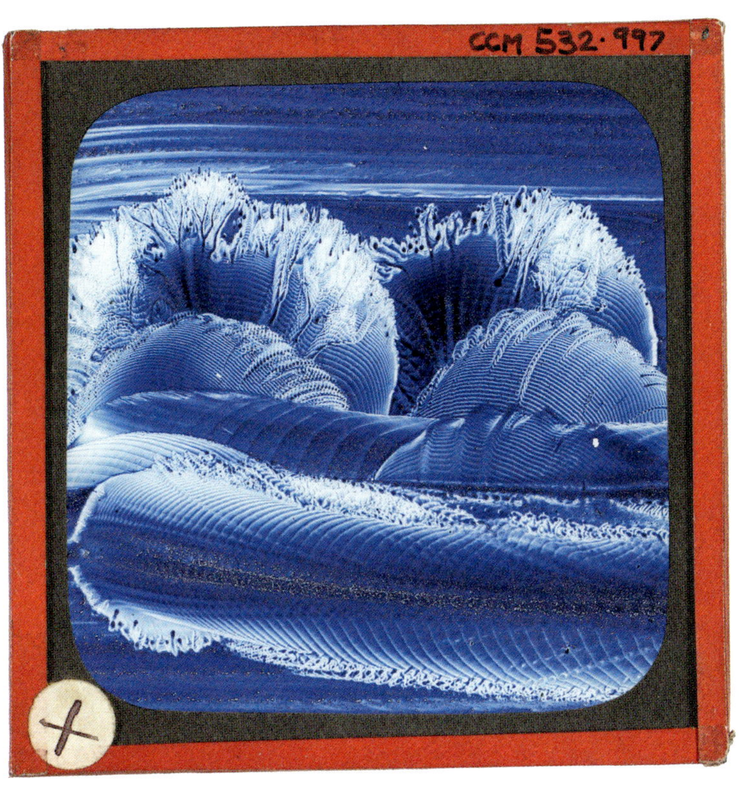
CCM 532·997

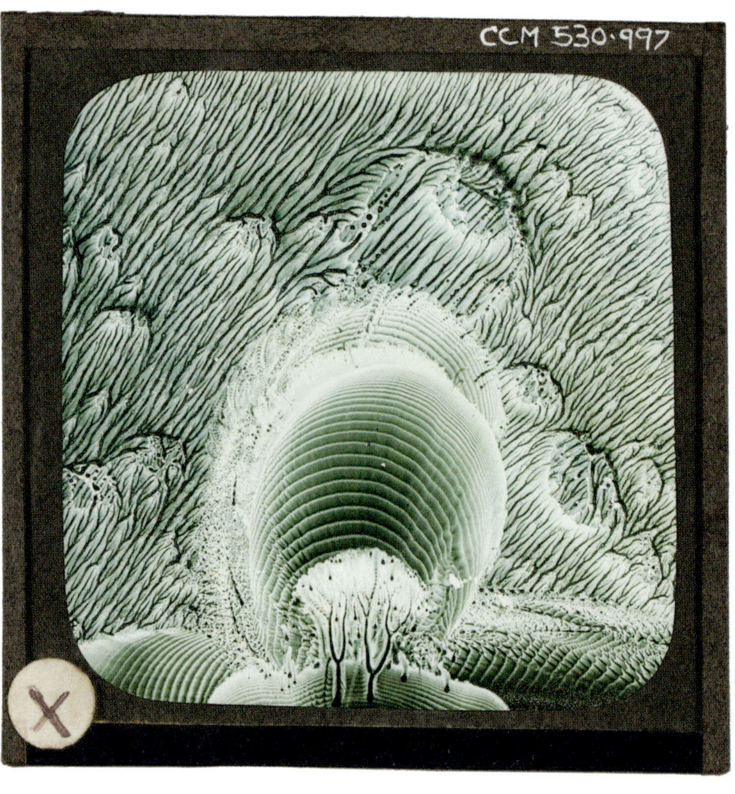

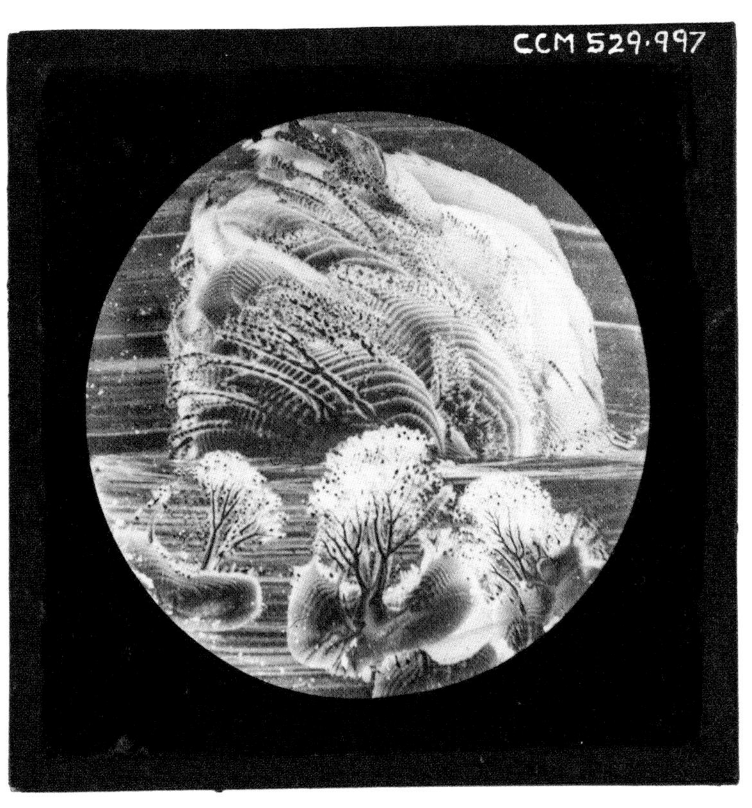

CCM 529·997

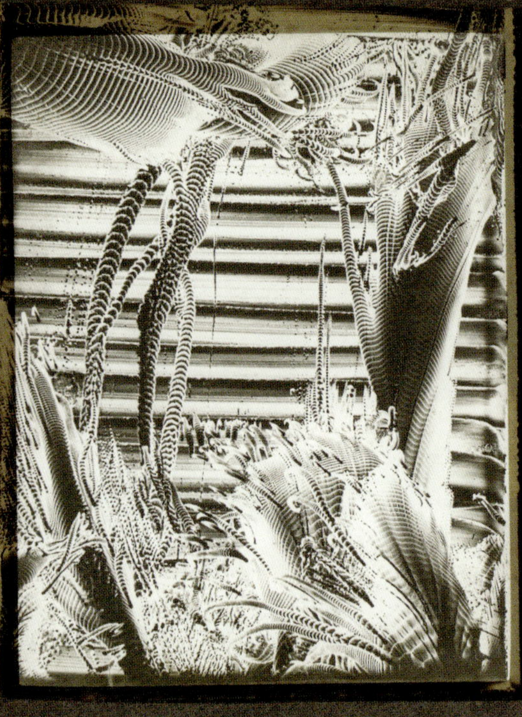
CCM 528·997

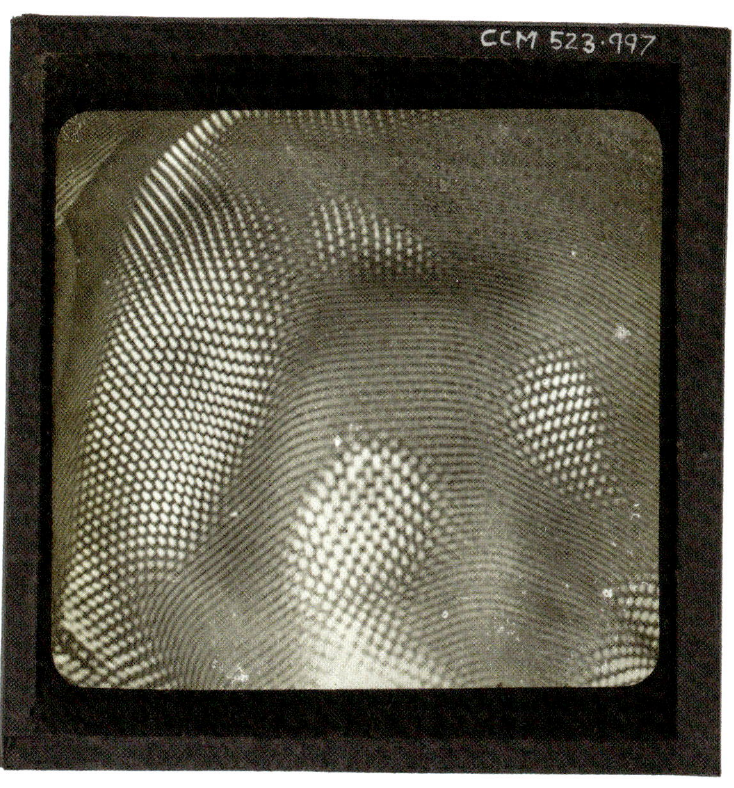
CCM 523·997

Margaret Watts Hughes and her invention of the eidophone were covered extensively in the international press upon publication of Voice Figures *in 1891. Interest was renewed in 1904 when she published the second edition,* The Eidophone: Voice Figures.

St. Louis Post-Dispatch, *May 10, 1908*

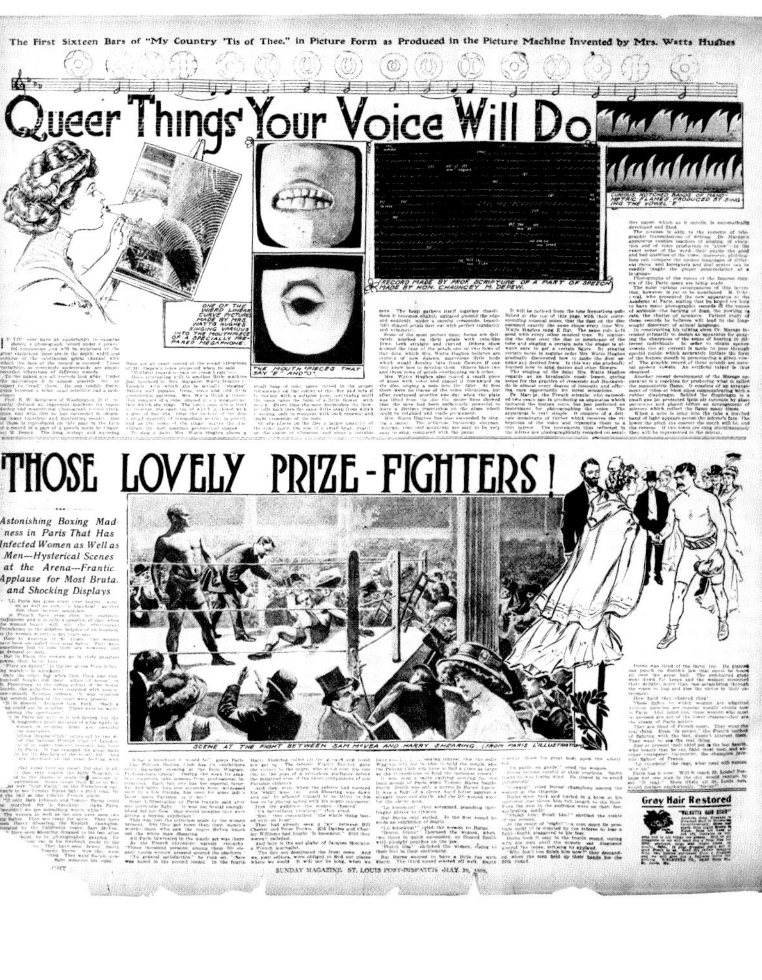

Queer Things Your Voice Will Do

THOSE LOVELY PRIZE-FIGHTERS!

Astonishing Boxing Madness in Paris That Has Infected Women as Well as Men---Hysterical Scenes at the Arena---Frantic Applause for Most Brutal and Shocking Displays

SCENE AT THE FIGHT BETWEEN SAM McVEA AND HARRY SHEARING. (FROM PARIS L'ILLUSTRATION)

Gray Hair Restored

"Can sound be seen? Why certainly, after a fashion, and it is very well worth looking at. These is an amazing series of illustrations of forms produced by sound accompanying a remarkable article in the May *Century* written by Mrs. Margaret Watts Hughes. Mrs. Hughes is a singer, but that she is also a woman of clear and original thought is proven by the account she gives of her surprising discoveries. She says that some years ago, while trying to find a way of indicating the intensities of vocal sounds, she met with "voice figures," or forms produced in fine powder or thin paste by the vibrations of the air produced by different notes. The apparatus used was merely a piece of sheet rubber stretched over the mouth of a small vessel, into which the voice was introduced by a tube with a wide mouthpiece. On this rubber membrane she spread a fine powder or thin paste or liquid. Upon singing notes of suitable pitch she found that beautiful crispations appeared upon the surface of the liquid which varied with every change of tone, and carrying the investigations further with colored glycerine, moistened powder and finally very small quantities of a colored paste she found that the voice produced forms closely like those of nature, the same note giving the same form and the slightest variation showing. For example, one of the illustrations shows in the centre a perfectly formed daisy. Another closely resembles a pansy; there are two that would pass for careful drawings of sea weed growing in the ocean bottom; one shows the shape of a tree, another of thick leaved ferns. There is a very remarkable serpent form, with many convolutions, the effect of perspective being perfect and the tail of the reptile seeming to project from the picture.

"When we know that the originals of these pictures were made without touch of hand and are the work of sound vibrations we seem to be on the verge of a great advance in human knowledge. . . .

"Is the violet the form of a sound, the sunflower a materialized musical note?"

The Lancaster Daily Intelligencer, *Pennsylvania, May 4, 1891,* *"Visible Sound"*

Daily Intelligencer.

ANDREW J. STEINMAN,
CRAIG STEINMAN & FOLEE, Editors.
ROBERT CLARK, Publisher.

THE DAILY INTELLIGENCER—Published every day in the year, but Sunday. Served by carriers in this city and surrounding towns at ten cents a week. By mail five dollars a year in advance; fifty cents a month.

THE WEEKLY INTELLIGENCER—One dollar and fifty cents a year, in advance.

NOTICE TO SUBSCRIBERS—Remit by check or postoffice order, and where neither can be had procure the presence and to a registered letter.

Entered at the Postoffice, as second class mail matter.

Address, THE INTELLIGENCER,
Lancaster, Pa.

LANCASTER, PA., May 4, 1891.

Let Both Run.

Our Republican friends are very evenly divided in their choice for judge and we suggest to them that they run both candidates; which will make another interesting contest before the people when the Democratic candidate's judge in the triangular race...

Visible Sound.

Can sound be seen? Why certainly, after a fashion, and it is very well worth looking at. There is no amusing series of illustrations of forms produced by sound accompanying a remarkable article in the May Century written by Mrs. Margaret Watts Hughes, Mrs. Hughes is a singer, but that she is also a woman of clear and original thought is proven by the account she gives of her surprising discoveries...

The Empty Treasury.

Pennsylvania has not yet received her direct tax payment and we incline to think that it may be some time before the will...

"Sing into a baking-powder can and a daisy will grow before your eyes. Sing again and a pansy will spring out. Sing again, and again and again, and the sand, or powder or paste on the rubber diaphragm which is stretched across the top of the can will tremble and shiver and form itself into circles, ovals, spirals and little mounds which seem alive with nervous action. The tin baking-powder can, with the rubber stretched across the top, is an 'eidophone' and with it sound becomes visible."

Daily Evening Item, *Lynn, Massachusetts, April 25, 1895, "Wonder in Vibration"*

WONDER IN VIBRATION.

How an Evening at Home Can Be Made Interesting.

Hitch a Can to a Toy Balloon and Watch Results.

Sound Waves Are Made Visible and Beautiful Forms Appear.

VOICE FIGURES MADE WITH THE EIDOPHONE.

HOME MADE EIDOPHONES.

ITEM EXTRA

FOUR O'CLOCK.

BY TELEGRAPH

UNCONSTITUTIONAL

Veteran Preference Act So Declared by Supreme Court.

Find That Legislature Cannot Make Such Law.

Case of Brown vs. Civil Service Commissioners.

"The laws which govern the dispersion, attraction, and cohesion of the particles suggest to her [Watts Hughes] the principle of universal integration and disintegration of matter. Readers of the holy scriptures who make a careful study of the voice figures, can but be struck with the analogies and the suggestions they contain. In all sacred books of the world the same thought is met. In the book of Genesis one reads 'God said . . . and it was so.' In St. John's gospel it is the word, 'without which was not anything made that hath been made.' The Hindoos speak of Vach, the creative voice; the Greeks say 'Logos'; the Romans, 'verbum,' each meaning word. Pythagoras said that he heard 'the ordered music of marching orbs.'

"According to occult science all this means that the electric vibratory energy which thrills through the cosmos, linking planets to their central sun as it links cell to cell, molecule to molecule, atom to atom, and even electron to electron, works under the same law of harmony that governs music. . . .

"Now, when the sun and planets of our system are drawn out to the scale of their space relationships they reproduce exactly the musical spacing of a fundamental note and its harmonics. This suggests that there is more than poetical imagery in 'the music of the spheres,' or the psalmist's allusion 'to the morning stars singing together.'"

The Chicago Sunday Tribune, *April 12, 1908, "How to Paint Flowers with Your Voice Instead of Your Brush"*

HOW TO PAINT FLOWERS WITH YOUR VOICE INSTEAD OF YOUR BRUSH

IT IS JUST AS EASY TO SING DAISIES AND PANSIES AS IT IS TO DRAW THEM. CONJURING INTO AIRY FIGURES EVERY MANNER OF GEOMETRICAL FORMS, SNOW CRYSTALS, FLOWERS, TREES AND OTHER THINGS, IS THE LATEST MUSICAL FAD.

"The Old Hundredth."

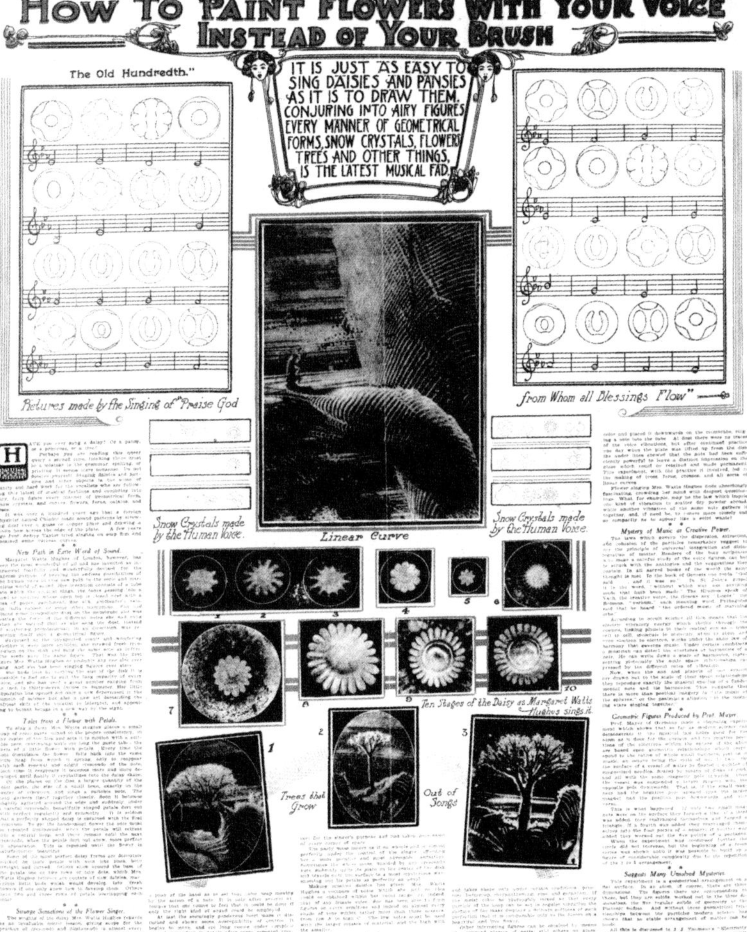

Pictures made by the Singing of "Praise God

from Whom all Blessings Flow"

Snow Crystals made by the Human Voice.

Linear Curve

Snow Crystals made by the Human Voice.

Ten Stages of the Daisy as Margaret Watts Hughes sings it.

Trees that Grow

Out of Songs

HAVE you ever sung a daisy? Or a pansy, or a primrose, or a tree?

Perhaps you are reading this query a second time, thinking there must be a mistake in the grammar, spelling, or printing it seems mere nonsense. Do not picture yourself singing daisies and pansies and other objects in the scene of sanity and hard work for the vocalists who are following this latest of musical fashions and conjuring into airy, fairy figures every manner of geometrical forms, snow crystals, and exotic flowers, ferns, cabbage, and trees.

It was over a hundred years ago that a foreign physicist named Chladni made sound patterns by strewing sand over a glass or copper plate and drawing a violin bow across the edge of the plate. A few years ago Prof. Sedley Taylor tried singing on soap film and obtained some curious effects.

New Path in Eerie Word of Sound.

Margaret Watts Hughes of London, however, has been the most wonderful of all and has invented an instrument carefully and wonderfully devised for the express purpose of proving the endless possibilities of the human voice in this new path in the eerie and mysterious world of sound. Her invention consists of a tube into which the vocalist sings, the voice passing into a ...

Makes Primroses and Other Flowers

Strange Sensations of the Flower Singer.

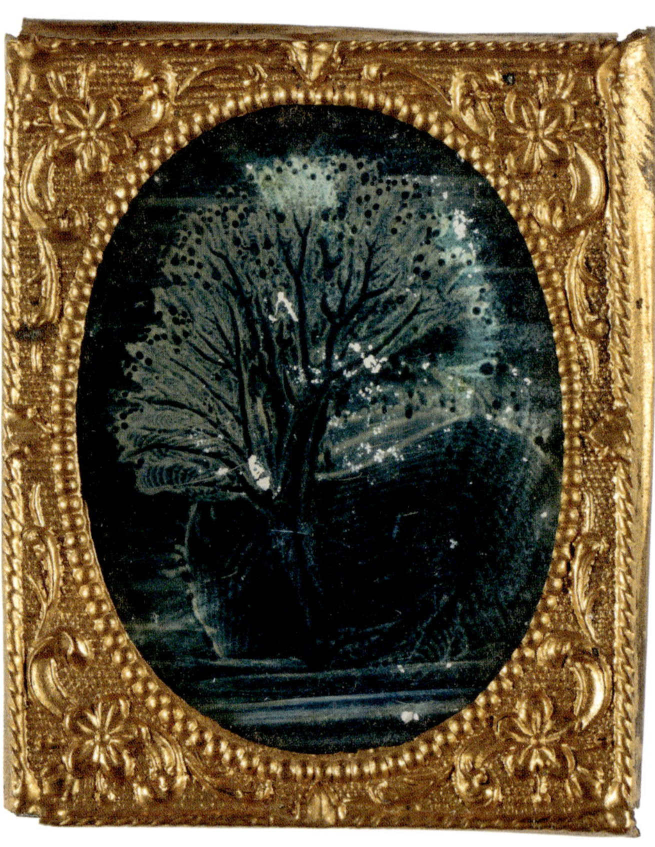

Besant, Annie, and C. W. Leadbeater. *Thought-Forms*. Theosophical Publishing Society, 1905.

Higgie, Jennifer. *The Other Side: A Story of Women in Art and the Spirit World*. Pegasus Books, 2024.

Mullender-Ross, Rob. "Picturing a Voice: Margaret Watts Hughes and the Eidophone," in *The Public Domain Review: Selected Essays* 7. PDR Press, 2019.

———. "Divine Agency: Bringing to Light the Voice Figures of Margaret Watts Hughes." *Sound Effects* 8, no. 1 (2019).

Voice Figure, ca. 1885–1904, paint on glass, 1⅞ × 1½ in. (4.8 × 3.8 cm)

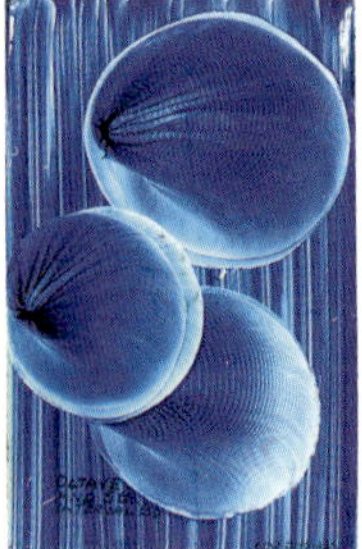
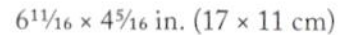

6¹¹⁄₁₆ × 4⁵⁄₁₆ in. (17 × 11 cm)

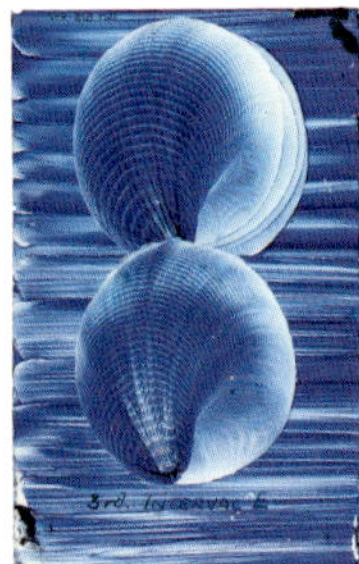

6¹¹⁄₁₆ × 4⁵⁄₁₆ in. (17 × 11 cm)

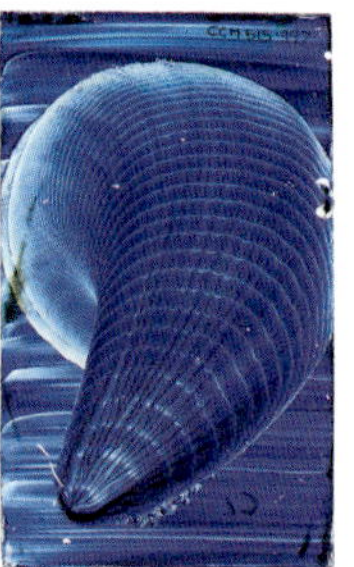

4¼ × 2¹¹⁄₁₆ in. (10.8 × 6.8 cm)

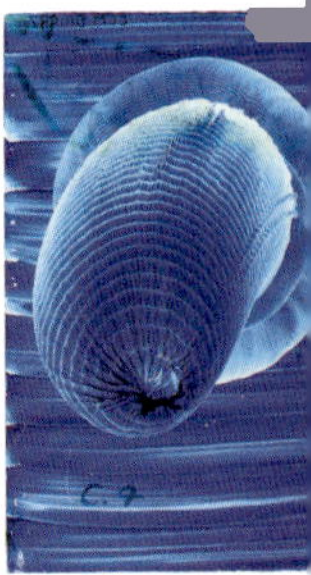

4¼ × 2¹¹⁄₁₆ in. (10.8 × 6.8 c

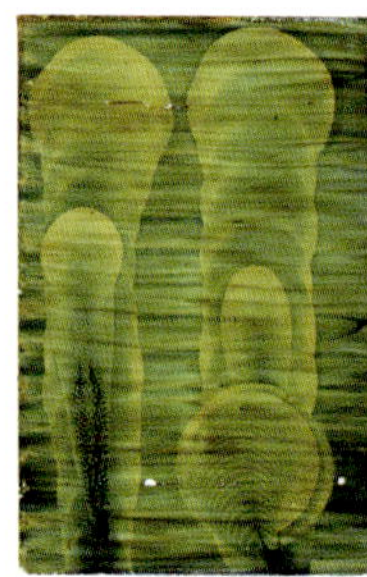

12 × 7⁵⁄₁₆ in. (30.5 × 20.1 cm)

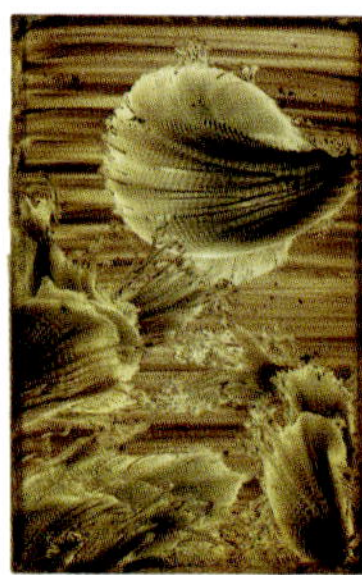

12 × 7¹⁵⁄₁₆ in. (30.5 × 20.2 cm)

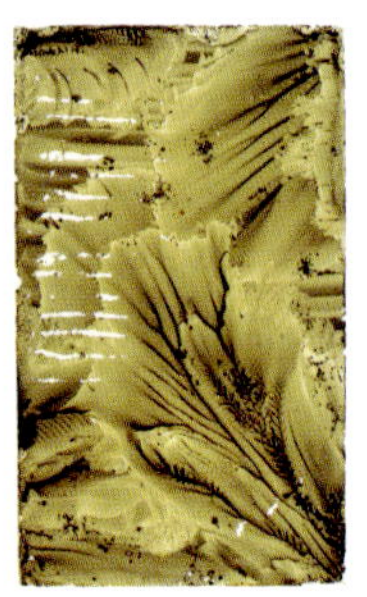

10½ × 6¾ in. (26.6 × 17.1 cm)

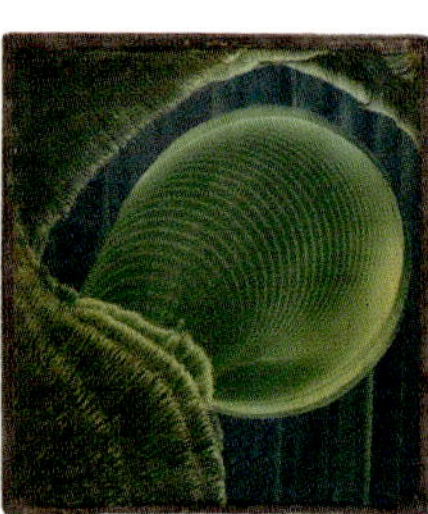

8¼ × 7⅝ in. (20.9 × 19.4 cm)

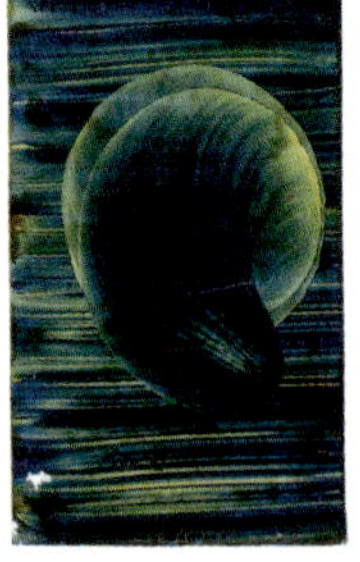

4¼ × 2¹¹⁄₁₆ in. (10.8 × 6.8 cm)

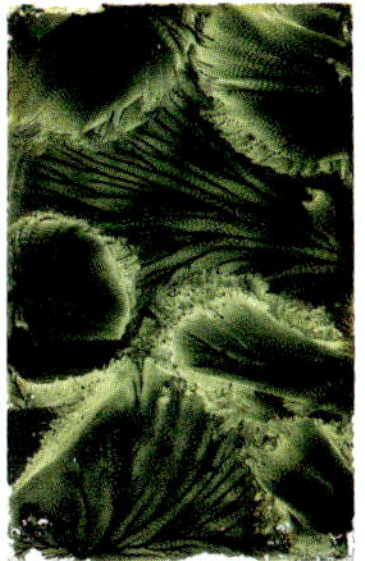 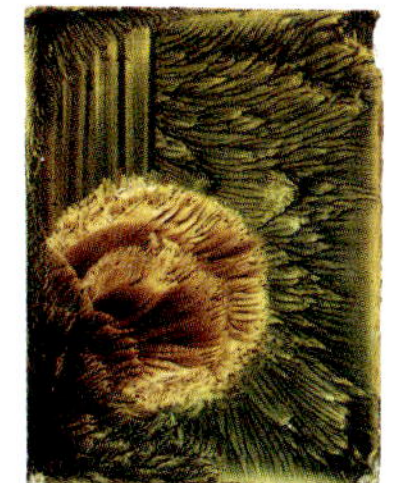

13 × 8⁷⁄₁₆ in. (33 × 21.4 cm) 9¹⁵⁄₁₆ × 7⅝ in. (25.2 × 19.3 cm)

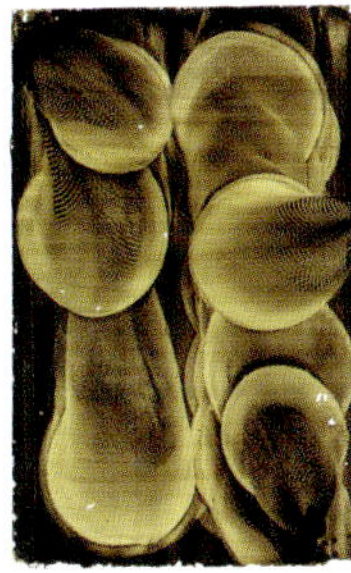 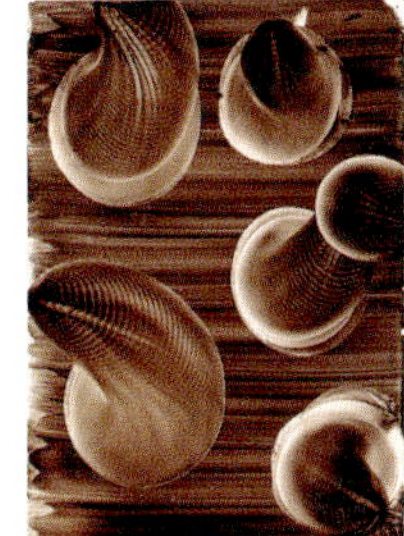

10 × 7¹¹⁄₁₆ in. (25.4 × 19.5 cm) 10 × 7 in. (25.4 × 17.8 cm)

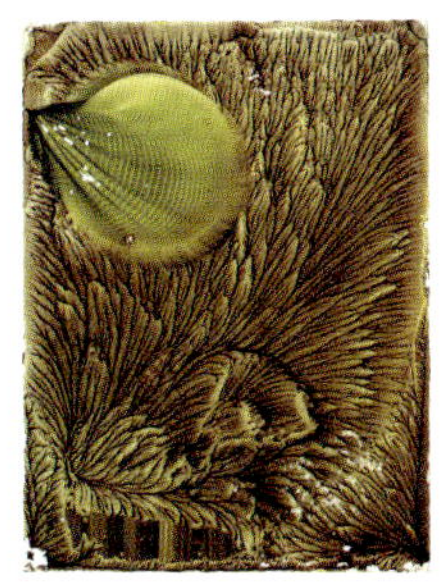 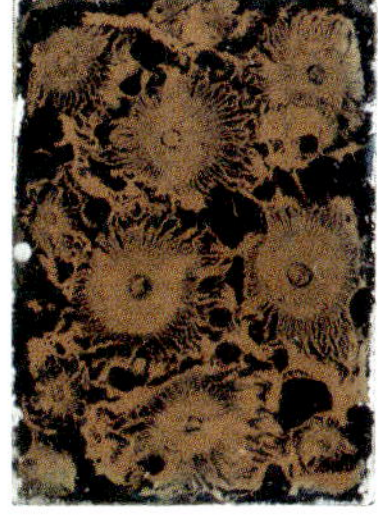 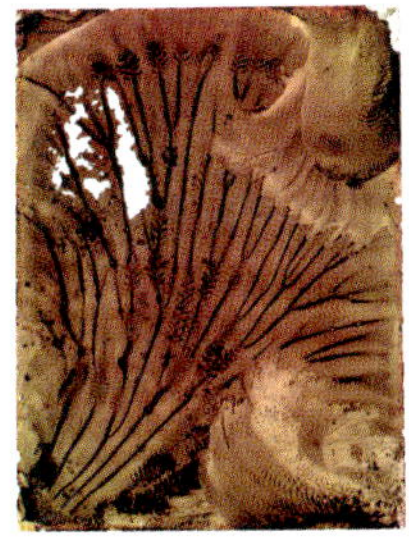

10 × 7¹¹⁄₁₆ in. (25.4 × 19.5 cm) 3½ × 2⁹⁄₁₆ in. (8.9 × 6.5 cm) 9⅞ × 7⅝ in. (25.1 × 19.4 cm)

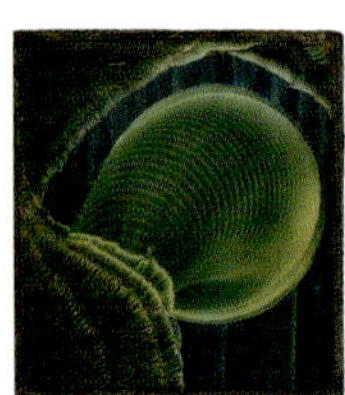

FRONT COVER

Voice Figure, ca. 1885–1904, paint on glass, 8¼ × 7⅝ in. (20.9 × 19.4 cm)

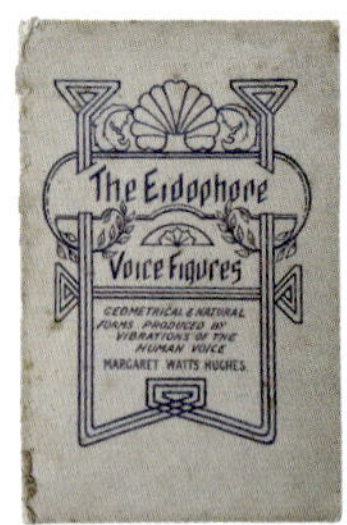

BACK COVER

Margaret Watts Hughes. The Edidophone: Voice Figures, *1904*